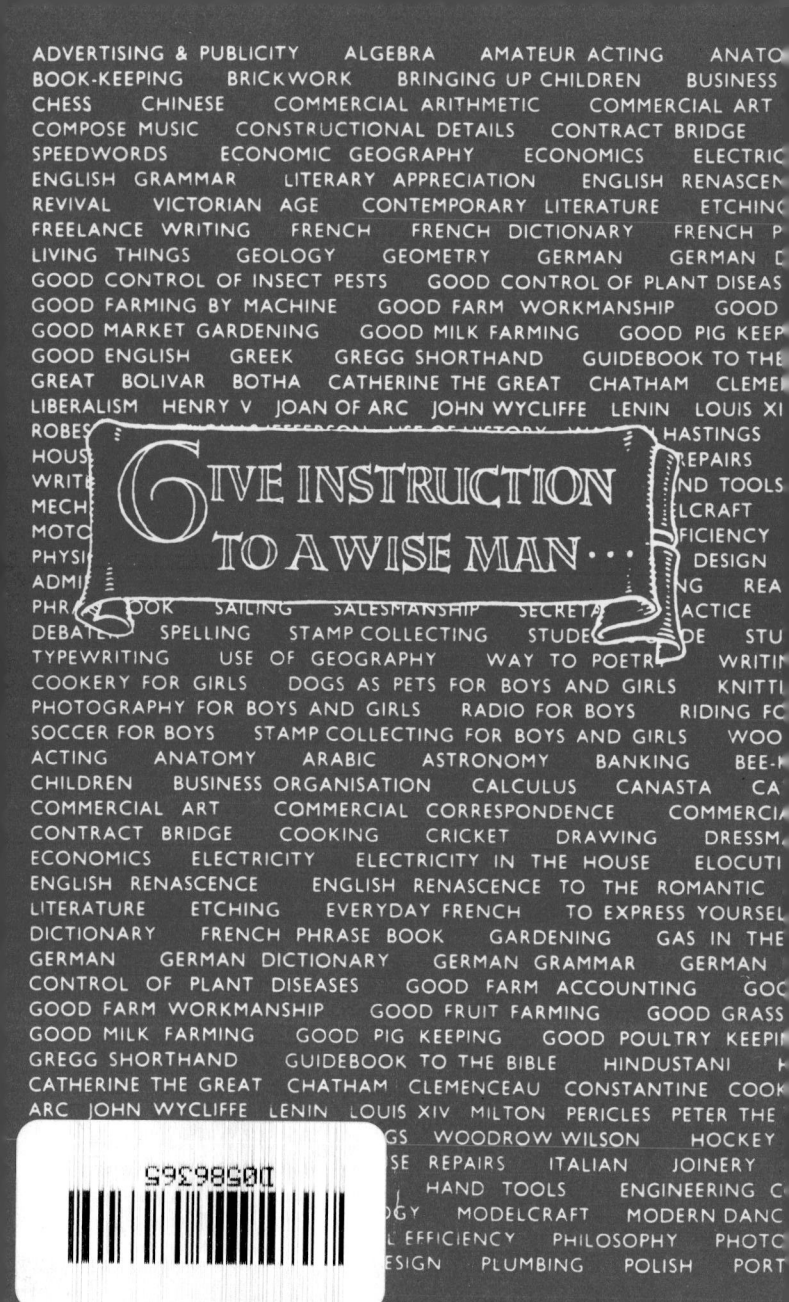

25p

THE TEACH YOURSELF BOOKS
EDITED BY LEONARD CUTTS

BEE-KEEPING

TEACH YOURSELF
BEE-KEEPING

By
LT.-COL. A. NORMAN SCHOFIELD, LL.M.
Solicitor

Illustrations by
W. W. NEWMAN, M.Inst.C.E., F.S.I.

Foreword by C. G. BUTLER, Ph.D.
of the Rothamsted Experimental Station, Harpenden.

THE ENGLISH UNIVERSITIES PRESS LTD.
102, NEWGATE STREET
LONDON, E.C.1.

First printed 1943
Revised Edition 1958

*Printed in Great Britain for the English Universities Press, Ltd.,
by C. Tinling & Co., Ltd., Liverpool, London and Prescot.*

FOREWORD

Bee Research Laboratory,
Rothamsted Experimental Station,
Harpenden, Herts.

MR. NORMAN SCHOFIELD has done me the honour of asking me to criticise the manuscript of this little book, and to write a short foreword.

In the past it has unfortunately been only too easy to criticise to their disadvantage many of the books on bee-keeping, since their authors often appeared to be ignorant of the advances in bee research, which had already been made at the time the books were written. What is worse, some of the authors of the past appear merely to have taken a few text-books on bee-keeping, and having selected what they considered to be best in each, to have copied it almost verbatim, thus perpetuating mistakes. Frequently, the results have been a mere re-hash, and often inferior ones, of earlier works.

Mr. Schofield has, I am glad to say, done no such thing; he has written an easily readable little book, based on his own practical experience as a bee-keeper, and I am convinced that the expert as well as the beginner will find something of value here.

I am very pleased to find that many of the old fallacies such as the idea that bees must be coddled with much packing in the winter, and that cane-sugar obtained from sugar-beet is harmful to bees, are conspicuous by their absence. Mr. Schofield will, I am sure, agree that the following words of Pope, which I recently saw quoted in a Bee Journal, give excellent advice to modern bee-keepers :

" Be not the first by whom the new is tried,
Nor yet the last to lay the old aside."

C. G. BUTLER

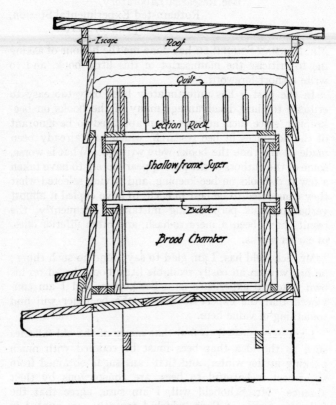

Escape

Roof

Quilt

Section Rack

Shallow frame Super

Excluder

Brood Chamber

THE W.B.C. HIVE

PREFACE TO FIRST EDITION

DURING the present war the value of sugar, and commodities containing sugar, have been more appreciated by the general public than in years of peace.

It is not surprising therefore that many thousands have turned to the most ancient means of producing sweetening material, namely, the keeping of the honey-bee. There are those who imagine that this is a simple matter to which they may readily turn a hand. Experience will show that a great deal of care is essential for, as in most practices, success will only be achieved by patience and perseverance.

There are those, too, who embark upon bee-keeping as a means to an end, namely, the production of honey for honey's sake, and who regard the bees as being the medium.

The true bee-keeper is the man who will see that his bees receive the best possible attention, who sees that they have always plenty of food, and who keeps them free from disease. He will reap his reward in much greater measure than the man who takes from the bees their last ounce of honey.

Unless the reader has no other object than the production of honey he should refrain from bee-keeping and leave these valuable creatures to those who will care for them.

It becomes an absorbing hobby to those who put a real interest in it, as do all studies of nature. It provides scope for the handyman. It can become a profitable hobby if due care is taken. It does not interfere with normal holidays and it can be carried on by the townsman just as well as the countryman, although the townsman has the disadvantage of travelling to his bees. Little ground is necessary, and only a small capital expenditure.

There is no more contentious subject matter amongst bee-keepers than bees, Many readers may disagree with many of my ideas. I should be grateful to have their views and the results of any experiments they may have carried out.

PREFACE TO THE FIRST EDITION

This book is intended to be an elementary guide to bee-keeping. Many of the finer practices are purposely omitted for fear of confusing the beginner. There are many advanced books on this matter to which the reader can readily refer—most public libraries contain a good number of them.

My sincerest thanks are due to Dr. C. G. Butler, of Rothamsted, for reading through the manuscript and for his suggestions and corrections. Also my thanks are due to Mr. Newman, also a bee-keeper, for his excellent drawings.

I must not forget to thank my good friend, Mr. F. Farquharson, of Watford, for all he has taught me.

A. NORMAN SCHOFIELD

The White House,
  Hempstead Road,
    Watford.
      Nov., 1943.

PREFACE TO THE THIRD EDITION

SINCE the publication of the earlier Edition I have received a great many letters from readers who have sent me their views on many matters. Certain errors in those Editions have now been rectified as a result of these letters.

This correspondence has been extremely helpful and I hope has improved this little book.

I refer to what is a comparatively new discovery in Chapter IV, dealing with the selection by the bees themselves of the sex of the resultant bee developed from the eggs laid by the queen. This may cause a great deal of controversy, but if in the end it leads to the truth, then it will have been worthwhile introducing it here.

Again I welcome observations and criticisms.

A. NORMAN SCHOFIELD.

Southampton,
  March, 1958.

TABLE OF CONTENTS

ix

4*

LIST OF ILLUSTRATIONS

PLATES

ILLUSTRATIONS IN TEXT

CHAPTER I

THE PURPOSES OF BEE-KEEPING

Honey

FROM time immemorial man has pursued the art of bee-keeping with the primary object of securing honey for food. Until the introduction of sugar derived from the sugar cane and sugar beet, it was the only known method of sweetening. To-day, the honey bee is used almost entirely as a producer of food.

As a food, honey has a very high market value and it is generally understood that it is one of the most easily assimilated of foodstuffs. Many ardent bee-keepers claim that it has virtues almost reaching the romantic. It is, however, used extensively in the making up of medicines, particularly for those prescribed for the throat, probably because of its syrup-like nature and the claim that it is an excellent antiseptic.

It is at all times a very welcome addition to the table.

Wax Production

A most valuable by-product of honey is beeswax. The story of its production is told elsewhere in this book. This product is of very great importance to the bee-keeper and it cannot be too strongly emphasised that not one scrap of it should be discarded by the bee-keeper. It should be stored in a tin so that the wax moth which is destructive of wax comb cannot destroy it. Later it should be melted down and clarified and sold to the manufacturers of wax foundation.

Shown later is a diagram of a solar wax extractor. Pieces of old comb are placed inside and the lid closed down. The sun's rays have a greenhouse effect and the heat generated melts the wax (see *Fig. 25*).

Natural History Study

For the naturalist and the student of natural history, the honey bee offers a fascinating study. Many people who commence bee-keeping for the purpose of producing honey soon become captivated with the study of the lives of the insects from which they reap such a sweet harvest. A great deal of knowledge has been recorded in recent years by bee-keepers, particularly since the introduction of the movable frame hive, and many of the mysteries to which ancient bee-keepers attached superstitions are normal and regular happenings in the hive.

Pollination of Blossoms

Apart from the production of honey, the pollination of blossoms is a most important role of the honey bee. Some experts, in particular Doctor Butler, of Rothamsted, assert that the production of honey is only secondary compared with the value of the honey bee as a pollinator of blossoms. Zander's original estimate was that the financial value of pollination was ten times the value of honey. It has now been established beyond all doubt that the yield of fruit from orchards having hives of bees exceeds many times those without. Whilst there are many, many insects, and indeed the elements of wind and rain which act as pollinators, none are so consistent and subject to such control as the honey bee. Orchard trees, however, are not pollinated by the wind at all. This has been shown by experiment at the John Innes Horticultural Institute.

Farmers and seedsmen have now come to realise the value of the honey bee as a pollinator of such crops as white clover, sanfoin, lucerne, the brassica family and the like. It has been discovered in the case of red clover that under favourable conditions the bees will visit it every third or fourth day for, on account of the shortness of their tongues they are unable to reach the nectar until it rises sufficiently in the flower for them to reach it.

Therefore, for the fruit grower, hives of bees are essential. It has been recommended that one hive to the acre of fruit trees is adequate to ensure complete pollination.

Breeding for Sale

This side of bee-keeping is the most commercial and at the present time very profitable. A stock of bees, with care, can be trebled or quadrupled in a season if honey is not worked for, and this will show a gross profit of from ten to fifteen pounds per hive, which is seldom, if ever, obtained from honey-producing hives.

CHAPTER II

THE COLONY OF BEES

A COLONY of bees consists of a hive containing combs, a Queen bee, a large number of worker bees and a number of drones.

The hive may be one of many types, but the beginner is well advised to purchase equipment of a standard type and to purchase all future supplies of the same standard. In this way, all his equipment will be inter-changeable, which is a most important thing. This cannot be too strongly emphasised. The standards which are in the commonest use in this country are British Bee-keepers Association Standards.

There are three principal kinds of hives in use in this country. (*a*) The straw skep hive, (*b*) the W.B.C. double-walled hive, and (*c*) the single-walled National hive.

The Straw Skep Hive

This hive has many advantages and many disadvantages over the modern movable frame hive, but the disadvantages far outweigh the advantages.

It is, however, cheap for it costs about ten shillings, it is light and easily moved from place to place. This was found convenient in the old bee-keeping days when transport was difficult and when bees were taken to the heather after the field flowers had died down. It produces swarms which are of value for stock increases. It can be "supered," a process which is later described. It certainly produces honey, for Pettigrew, in his "Handy Book of Bees," 1875 records that a Mr. Jack, of Wilshaw, had a straw hive and its swarms which, at the end of the season, weighed 474 lbs. A conservative estimate of the honey which could have been obtained from those straw hives was 300 lbs. Many modern bar-frame hive operators may wonder at these results, but these and others are recorded in that book.

Straw hives are useful when the bees are required for pollination only.

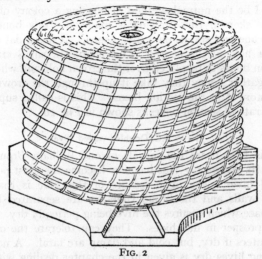

FIG. 2
The Straw Skep Hive

The colony which has become established in a straw skep may be transferred to a bar frame hive by placing the skep on top of a standard brood chamber as shown in Plate I. The queen will ultimately lay in the new brood chamber in the bars fitted there. When she is found " below " a queen excluder (see page 27 and *Fig.16*) should be put between the skep and the brood chamber.

The brood chamber is that part of the hive where the queen lays her eggs and the young brood are reared (see *Figs. 4* and *5*).

Straw hives, **however,** do not permit of complete control in management. They may harbour disease and be a danger to other stocks. If, however, the bee-keeper is bent upon a complete study of his subject, he should keep one stock in a straw hive in his apiary. On account, however, of the present serious position with regard to bee diseases, in particular foul brood, it is not recommended that the skep should be the normal method of housing a colony of bees.

The skep in this drawing is made of straw bound by split cane and stands on a hive board. It has a flat crown and a hole in the centre. This enables a small crate of section boxes to be placed over the crown hole and when the bees gather honey they take it up through the crown hole and store it in the section boxes. This is called supering.

A crate of section boxes is shown in *Fig. 26*.

The W.B.C. Doubled-walled Hive

This hive bears the initials of one *W*illiam *B*roughton *C*arr, its inventor. This has now become the most popular of all hives. The main feature of this hive is that the outer walls and the brood chamber are separated by an air space which ensures the hive being perfectly dry. Bees only prosper in dry hives. They can tolerate the coldest of winters if dry, but cold and damp are fatal. A note on keeping hives dry is given in the chapter dealing with the wintering of bees.

From the diagram of this hive (ante, at page vi), it will be noticed that the brood chamber is placed on a stoutly constructed wooden stand. On the stand outside the brood chamber is placed the outer wall, this is built up in tiers, the bottom of one fitting on top of, but on the outer side of the one below. These tiers are called " lifts." These lifts fit on each other like slates or tiles to prevent the rain from getting inside. The whole is covered in with a roof which either slopes from the centre outwards or from front to back. The latter is preferred as it ensures the rain being kept clear of the entrance to the hive. It is also easier to make and has a very smart appearance (see also *Fig. 3* showing the outward appearance of a W.B.C. hive).

The brood chamber is of B.B.K.A. standard measurements and is designed to hold 10 B.B.K.A. standard brood frames and a division board. This board is a desirable fitting, as it ensures that all the standard frames fit snugly together. The frames hold the wax foundation and are to be seen in *Figs. 6* and *10*. If the frames do not fit correctly together, the bees build comb between the frames and at right-angles to the frames. This is called brace comb and

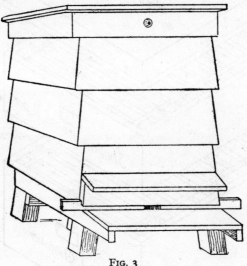

Fig. 3
The Outer Structure of a W.B.C. Hive

is a great nuisance when manipulation of the hive is in progress. Further, if the combs are not set snugly together, the bees are apt to build drone comb in the frames. When manipulating a hive, this board is first taken out and the space left in the brood chamber enables the frames to be eased and lifted out more readily and without crushing any bees.

In the diagram of the W.B.C. hive (frontispiece) will

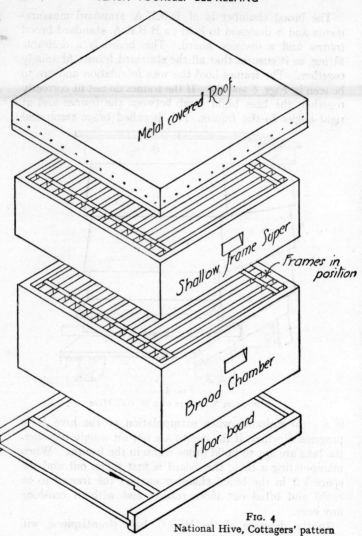

FIG. 4
National Hive, Cottagers' pattern

be noticed another chamber over the brood chamber. This chamber is constructed so as to hold a number of shallow frames in which is stored the honey. This chamber or crate is called a " super " or shallow frame crate.

In place of this super or in addition to it may be placed a crate of sections if the production of section honey is required (see *Fig. 26*).

Again, it should be emphasised that if the beginner decides upon the W.B.C. hive, he should ensure that all the outside lifts are interchangeable.

The National Hive

The characteristic of this hive is that it is single walled only. That is, the walls of the brood chamber form the only protection between the bees and the outside atmosphere. It has proved to be perfectly satisfactory, provided that the hive is kept dry. It is cheaper than the W.B.C. hive, both initially and in the long run, and it is more economical than the double-walled types. The National hive holds eleven brood frames in its brood chamber, it has less parts than the W.B.C. and is more readily portable, and for this reason, is ideal for use for those bee-keepers who place their bees on the heather.

The single-walled hive can be purchased in a cheaper form known as the Cottagers' hive (*Fig. 4*). This hive is placed on a stand. A wide drain pipe set into the ground is most suitable, as its smooth surface prevents mice and other pests from entering the hive. Its great disadvantage is the risk of damp entering the hive.

Fig. 5 shows a National Hive with a deep cover. In winter time when the super is off it acts as a double wall to the brood chamber.

The Brood Frame

As has already been described, the brood chamber in B.B.K.A. standard has 10 brood frames. *Fig. 6* shows a standard brood frame.

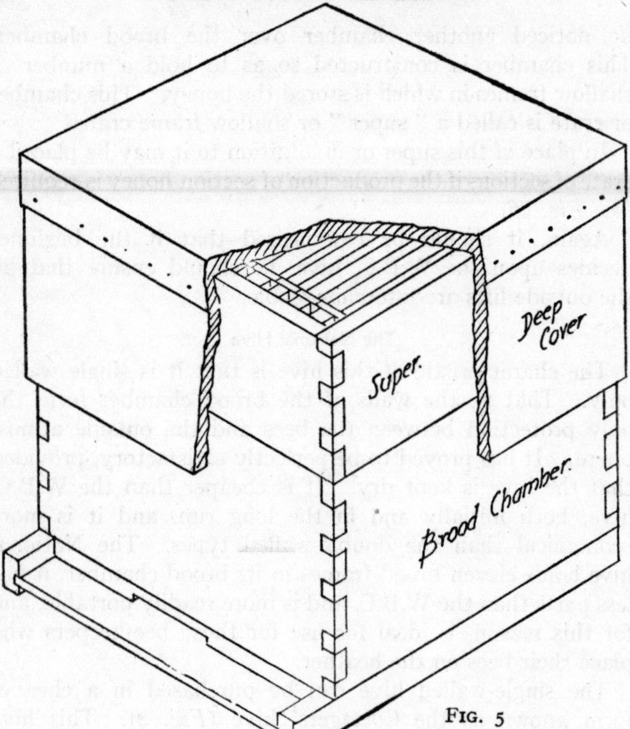

an oblong wooden chamber over the brood chamber. This chamber is constructed so as to hold a number of shallow frames in which is stored the honey. This chamber or unit is called a "super," or "shallow framed super." In place of this super it is common to find may be placed a shallow super to contain the production of section honey...

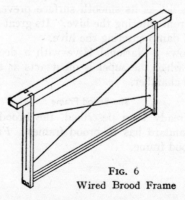

FIG. 5
National Hive with Deep Cover

FIG. 6
Wired Brood Frame

The measurement of the top bar is 17 ins. long, $\frac{7}{8}$ in. wide, and $\frac{3}{8}$ in. thick.

The side bars have an overall length of $8\frac{1}{2}$ ins. and the bottom bar has an overall length of 14 ins. Both side and bottom bars are $\frac{7}{8}$ in. wide and $\frac{3}{4}$ in. thick.

At the ends of the top bars, " metal ends " are fitted.

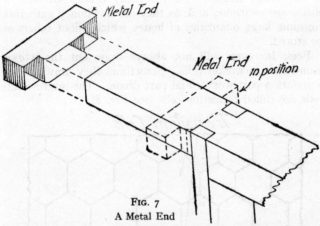

FIG. 7
A Metal End

These metal ends are designed to ensure that the frames are accurately spaced, for the bees build their brood comb $\frac{7}{8}$ in. thick from face to face with an intervening space between of $\frac{5}{8}$ in. This makes the metal end $1\frac{1}{2}$ ins. wide. Wider metal ends, 2 ins. wide, are sometimes fitted to standard shallow frames used for the production of extracted honey, and the use of these is described in the chapter on honey production.

Foundation

The invention of this material manufactured from pure beeswax is one of the greatest boons to bee-keepers. This material consists of a thin sheet of beeswax impressed mechanically with the forms of the bases of the cells of honeycomb and the bases of the cell walls. This is sold in sheets of the correct sizes to fit in the wooden brood frames, shallow frames and section boxes. For brood frames, " worker foundation " is supplied. This ensures that the bees are induced to build worker cells in the brood chamber to the exclusion of drone cells. It should here be pointed out that it is a great disadvantage to have too many drone cells in the brood chamber for too many drones are said to encourage swarming and, as they are not honey gatherers, consume large quantities of honey which might otherwise be stored.

Bees, however, will not always draw out the worker foundation as worker cells. Sometimes they draw out the foundation part worker and part drone. The intermediate cells are called transition cells (see *Fig. 8*).

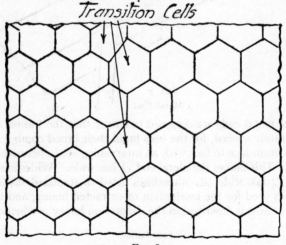

FIG. 8
Transition Cells

For shallow frames " drone foundation " is recommended, because the bees do not often store pollen in drone cells (see Plate 2, page 35), but honey only, and when honey is stored in drone cells it is more readily extracted than from worker cells. In natural surroundings bees build drone cells in which to store honey. It should be pointed out, however, that when drone foundation is used in the supers a queen excluder is absolutely essential. There are other disadvantages in the use of drone combs. Sometimes the bees hesitate to go into drone combs until the queen has laid eggs in them. This is frustrated if a queen excluder is used. Further they cannot be used as food storage combs for winter. There are those who think

The Correct Way   The Incorrect Way

FIG. 9

The Correct and Incorrect Way of Fixing Foundation

that the use of a queen excluder outweighs the advantages of drone cells for the storage of honey. It certainly is a great help to have drawn out worker combs in shallow frames to form an addition to the brood nest. On the whole drone combs for supers are preferable.

There are approximately 28·87 worker cells to the square inch ; each cell being approximately 1/5 in. in diameter. The cells are hexagonal in shape, two sides of the cells are vertical. This is important to remember for if the foundation is inserted with two sides horizontal, the bees will break down the foundation and build more drone cells than the bee-keeper desires. This causes delay and waste of effort on the part of the bees. *Fig. 9*, ante, shows the correct way up of foundation and also the incorrect way.

Drone cells are larger than worker cells, being approximately ¾ in. diameter. There are 18·48 cells to the square inch.

Foundation should be made from pure beeswax. Any

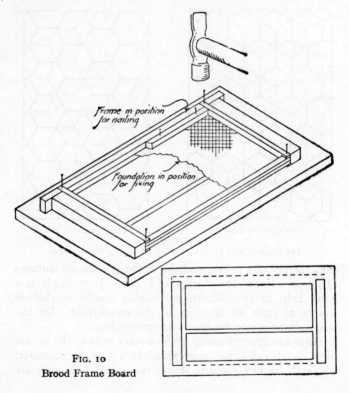

Frame in position for nailing

Foundation in position for fixing

FIG. 10
Brood Frame Board

foundation which is adulterated with any other material should be refused by the bee-keeper.

Foundation used for sections is made of the finest selected beeswax and is much thinner than for brood or shallow frame foundation. It is cut in squares 4⅜ in. each way, ready to fit into the sections.

Mr. Newman has devised a special board for ensuring that the frames when nailed together are absolutely square. The beginner would be wise to make one of such boards (see *Fig. 10*).

Queen Excluder

This is the name of a device which is placed over the brood chamber and which permits the worker bees to pass through but not the Queen. The Queen, being larger in the thorax than the workers, is therefore kept " downstairs " in the brood chamber, and does not lay any eggs in the supers. This is particularly desirable where sections are being worked—for in that case the honey must not be contaminated by breeding in the honeycomb which is to be eaten.

The excluders are made either of zinc with slots cut in it, as shown in *Fig. 16*, or are made of wire held firmly by separators. (*Fig. 17*.) The principal pattern in use is called the Waldron excluder, and this pattern is generally thought to be a very great advance on the zinc types.

The excluder is laid on the top of the brood frames, the slots in the excluder being laid at right-angles to the brood frames. The excluder covers the whole of the brood frames. Over the excluder the shallow frame crate or the section crate is set and over the crate the quilt.

It is essential for the bee-keeper to take great care with queen excluders. If the zinc type is used care should be taken to ensure that the zinc separating the slots does not become broken or buckled otherwise the queen will find her way through. Also with the Waldron excluder care should be taken to ensure that none of the wires are bent or the queen will also get through.

The Quilt

This is a piece of fabric. Some people use ticking or linen, but deck chair canvas or sail cloth proves as satisfactory as any.

Over the quilt is placed one or two dry clean sacks or one or two thicknesses of carpet under-felting.

Many bee-keepers prefer to use a clearer board all the year round. (*Fig. 20.*) The author himself has tried both canvas and clearer boards and much prefers the latter. In winter the clearer board dispenses with the necessity for bee passages referred to in Chapter X.

PURCHASING BEES

Purchasing a Stock

The beginner is advised to take the advice of any experienced bee-keeper in selecting a stock of bees.

The best type of stock to buy is an established stock on 10 brood frames, with a Queen born the year before. A Queen is thought to be at its best in the second year. The purchaser should insist that the stock is to B.B.K.A. standard.

The standard is as follows :—

Two-thirds of the number of combs in the colony should contain brood.

All combs to be well covered with bees and in each stock a fertile Queen must be present.

To purchase a stock is the most expensive method, but if honey is required in the first year, this is the surest way of obtaining it. It costs about £6-0-0.

The other ways of starting are as follows :—

Purchasing a Swarm

The advantages of purchasing a swarm are :—

(*a*) It is comparatively cheap. It costs about £3-0-0.

(b) If fed with warm syrup, the swarm becomes quickly established.

(c) There is an uncanny " urge " in a swarm to establish itself rapidly and produce a surplus for winter stores. If it is an early swarm, i.e., a May or early June swarm, a surplus is assured if the season is good.

The disadvantages of purchasing a swarm are :—

(a) It is always problematical how a swarm will develop. The Queen may be past her prime and the stock prove useless as honey producers.

(b) If the swarm is from an unknown source, the characteristics of the bees are uncertain and may prove savage or diseased.

Purchasing a Nucleus

A nucleus is a small stock of bees having a fertile Queen of the current year. The advantages of this way are :—

(a) This is a cheaper way than purchasing a full stock.

(b) The purchaser is assured of a young Queen.

(c) The purchaser knows the source and characteristics of his bees.

The disadvantages are :—

(a) The improbability of obtaining surplus honey in the current year.

(b) The cost of feeding up the nucleus to enable it to live through the winter.

The best advice is to purchase a full stock on at least eight frames in the month of May. In this way, with careful treatment in swarm prevention, the bee-keeper is assured of a surplus of honey in his first year.

CHAPTER III

HOW TO WORK FOR SURPLUS HONEY

IT must be impressed upon the beginner that the worker bees produce honey and that the more bees there are the more honey is likely to be produced.

Spring Stimulation

The first step to take to ensure a successful season is to winter the stock well and carefully. Then Spring comes and early Spring sunshine causes the bees to fly. When this time arrives it is the signal for the bee-keeper to take the first step in a definite plan.

In the last week in March in the South of England, and in the second or third weeks of April in the North of England, Spring stimulation should be started.

Spring stimulation consists of two matters :
(a) Stimulative feeding.
(b) Brood spreading.

There are many views in regard to feeding. Some say that a cake of candy placed over the feed hole and constantly renewed, is all that is necessary in the Spring. That certainly is an easy and troubleless method but it is not the most effective.

An established method and one which has proved very satisfactory is by giving the bees a cupful of warm thin syrup through the slow feeder. The slow or bottle feeder is shown in this diagram.

This dose should be given every second day. If too much syrup is given, the bees will spend their energies storing the syrup. The idea in giving them this syrup is to make them believe the honey flow has started. They then start feeding the Queen, who lays eggs in proportion to the amount of feeding she receives. The feeding should be kept up until the early blossoms appear and provide the nectar to take the place of the artificial food.

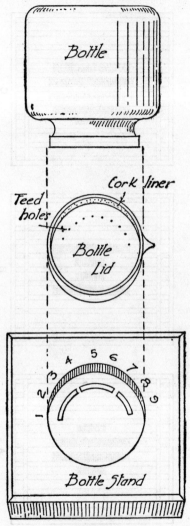

Bottle

Cork liner

Feed holes

Bottle Lid

Bottle Stand

FIG. 11
Slow or Bottle Feeder

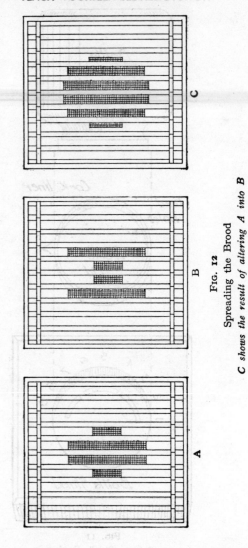

Fig. 12
Spreading the Brood
C shows the result of altering A into B

The usual accepted formula for Spring syrup is 1 pint of water to 1 lb. of sugar. Some give as a suitable recipe 2 pints of water to 1 lb. of sugar.

Dr. Butler, of Rothamsted, has recently been carrying out experiments on very thin syrup indeed, and it is understood with favourable results. The author, hearing Dr. Butler on this matter, tried feeding by a rapid feeder a pint of warm syrup daily of 1 lb. of sugar to 10 pints of water. The stock did extremely well, producing 92 lbs. of surplus honey in the 1942 season which, in the Home Counties, was not regarded as a very good season.

However, it should be clearly understood that too much sugar may prevent the Queen from laying because of the consequent filling up of the cells.

It is known that bees require a considerable amount of water when they are breeding, and feeding very dilute syrup to them saves them many chilly journeys in search of water in the early Spring mornings.

It is also known that the length of the life of a bee bears a direct reverse ratio to the amount of flying time. Therefore, if you can save bees a lot of flying time by feeding water to them, you will ultimately obtain more nectar collecting from your bee population.

On a warm sunny day about seven days after you start this thin syrup feeding, take a look at your bees. You will probably find a certain amount of sealed brood but a larger proportion of grubs and eggs. If you find large patches of eggs and grubs, you may be quite satisfied that your stimulative feeding is having effect.

Spreading the Brood

Having stimulated the Queen to activity with satisfactory results, it is now possible to extend the activities of the Queen. There are two conditions precedent to brood spreading and they must be adhered to very rigidly; otherwise, it may prove disastrous.

B

First, the weather must not be too cold or raw. Preferably, the third week in April must be reached.

Secondly, the hive must contain a large number of young bees to ensure good cover for the combs.

These two conditions are important and must be observed ; otherwise, the brood may become chilled and die.

The Queen in Spring lays her eggs in the centre of the cluster of bees.

It will be noticed from the diagram (brood chamber " A ") that the two centre combs have more brood in them than the combs adjoining them on either side. Now place the outer combs in the centre, as in brood chamber " B," and the effect is to encourage the Queen to lay in this fresh accommodation, resulting in brood chamber " C." After a week, move the other outer combs to the middle and a similar stimulation will be given to the Queen. As the bees hatch out, the brood will spread even further. Keep on bringing in the outer combs if the weather continues fair and warm, until all the combs are filled with brood. Do not get over enthusiastic too early in the season with this hazardous method of stimulation and on no account should any combs which have no brood in them be put in the centre of the brood until at least the second week in May. If this method of stimulation proves successful, be sure to give the bees plenty of space for storing honey, otherwise the stock will be so strong that it will swarm at the earliest opportunity. The other methods of swarm control should also be carried out. These are enumerated later. If the beginner is in any doubt he should avoid this practice and the risks it entails.

Spring Cleaning

Whilst all this Spring stimulation is proceeding, an opportunity should be taken on a warm fine day to clean up the interior of the hive.

One very good way of doing this is to prepare a spare

PLATE I.

Transferring an established stock of bees from skep to standard brood-chamber. The Queen when found in brood-combs should be prevented from going back into skep by a queen excluder.

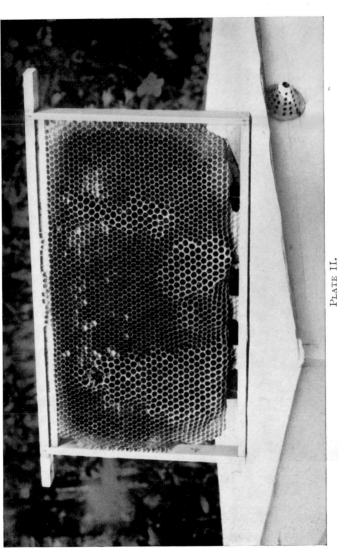

PLATE II.

A bad brood-comb. Combs like this should be taken from the hive, melted down and fitted with worker foundation. Note pollen in worker cells but not in drone cells.

hive and brood chamber and move the stock to one side and place the spare hive and brood chamber exactly where the stock previously stood. Then take out each brood frame in turn, examine each carefully for the effects of wax moth, or excessive amounts of drone cells in the combs. Scrape each frame clean of all unnecessary comb and propolis. Smear the metal runners in the brood chamber with vaseline and then place each frame in the new brood chamber in exactly the same order as in the old chamber. Take away the old hive and brood chamber and clean them up ready to deal with the next stock. Remove any combs which have an excessive number of drone cells. You want workers not drones (see Plate II).

Reference is made here to the use of vaseline. This is most desirable for it prevents the bees from cementing the metal ends together with propolis. It will be found later that any bar can be lifted out without the necessity of using a hive tool, chisel or screw-driver for loosening the bars. The absence of " jarring " has a marked effect on the bees and they are much more docile when vaseline is used.

Cleanliness in a hive is very important and it is a great help in resisting bee diseases. The " dirty " bee-keeper loses the respect of his fellow bee-keepers who come to regard his apiary as a source of danger to the district.

Propolis is described at page 48.

Capacity of Brood Chamber

As already stated, the B.B.K.A. standard brood chamber has ten standard brood frames, but this is considered by many to be too small. Experience has generally shown this to be so in many districts and with the best and modern strains of bees.

If Queens are prolific breeders it is estimated that they may lay anything up to 3,000 eggs per day, and as one brood comb has approximately 5,000 cells, i.e., including both sides of the comb, there is approximately one and a half days work in each comb, bearing in mind that not all the cells on each side of the comb are filled with eggs. As has been stated elsewhere, worker bees take 21 days from egg to bee—one and a half into 21 goes 14 times. Therefore, at the height of the season, 14 frames would appear to be required.

Many bee-keepers use two brood chambers of 10 brood frames in each, one placed above the other, but a brood chamber together with a crate of shallow frames having worker foundation proves most suitable. It is found that a normal Queen will keep these combs fully occupied with brood rearing.

There is a rational argument against too large a brood chamber and it is that too much effort can be wasted by the bees in rearing so large a stock to the detriment of honey production.

It is certainly true that at the height of the honey flow, if there are two or three cold days, large stocks consume an enormous amount of stored honey.

Many bee-keepers who use a brood chamber and shallow frames for breeding, discard the Queen excluder completely, especially when sections are being worked. But if this is done, drone foundation should not be used for the upper shallow frames in case the Queen should find her way up there and fill the drone cells with eggs. Queens do not particularly like section boxes for breeding and they remain below in the more spacious apartments. The absence of a Queen excluder is certainly an advantage to the worker bees, who very often refuse to work sections if there is a Queen excluder below the section crate.

Having now developed through stimulative feeding a strong stock, the bee-keeper must take very good care that the strength of his stock does not defeat his object.

Honey Storage by Bees

If there is not sufficient room for honey storage, the bees will follow their natural instinct and swarm. A separate chapter deals with the question of swarm prevention, but it is desirable to set out here briefly the ways to prevent swarming :—

(a) Having a young Queen.

(b) Have sufficient space for breeding.

(c) Have sufficient space for honey storage.

(d) Eliminate as many drones as possible.

(e) Inspect periodically, at least every 10 days, to remove any Queen cells, and, if necessary, give more room for honey storage.

Therefore, as soon as the brood combs appear to have their upper cells drawn out with new white wax, the bee-keeper must without delay give more room for honey storage.

Remove the quilt and place the Queen excluder on top of the brood chamber or, if brood chamber and shallow combs are used for breeding, then on top of the shallow combs. Over the Queen excluder place the new crate of shallow combs and over the shallow combs place the quilt.

Sometimes, it is found that the bees are somewhat slow in making use of these combs, particularly if they are frames with new foundation. The bees can readily be encouraged by placing a feeder of warm syrup over the feed hole in the quilt, or by splashing the new combs with a liberal quantity of sugar syrup. The bees will then rise very quickly and use the syrup to make the wax needed to draw out the foundation.

In many districts where the early fruit blossoms are in abundance, it is not unusual to get a little surplus honey provided the stock is a strong one and the combs are already drawn out.

Where foundation is used instead of combs, it will be found that those in the centre of the crate are drawn out first. It is a great advantage to have all the combs drawn out in readiness for the main honey flow. The bee-keeper is therefore advised to watch the combs and as the centre ones are drawn out, to move them outwards and place the outside combs in the middle and so on until all the combs are fully drawn out.

Before all the combs are drawn out, a considerable quantity of uncapped honey is found to be stored. If the weather continues favourable, the bees will begin to cap over the honey. They will not, however, cap over the honey until it is of the right consistency, that is, in such a state that it will not ferment.

As soon as one-third of the total surface of supers is capped over, the bee-keeper must place another super on the hive ; otherwise, the bees may swarm. This is the way in which the super should be placed on the hive. Take off the partly filled super and place the new super on top of the Queen excluder and then place the partly-filled super on top of the empty one. The reason for this is that bees naturally build their combs from the roof downwards, and will continue from the top super downwards to the brood chamber.

The bees in their travels up to the top super become accustomed to the new super and are soon seen preparing the combs to receive their store of honey. So long as the bees go on storing honey, so the bee-keeper should continue adding supers.

Some people say it is desirable to leave the honey on the hive until the end of the honey flow. This is not advised when sections are being worked because even though the combs are filled and sealed, the bees continue to walk over the surface of the combs and they become discoloured or "travel-stained," and this makes them less appetising and of a reduced market value.

CHAPTER IV

THE OCCUPIERS OF THE HIVE

At the height of the season, that is in the months of June and July, the bee-hive contains a very heavy population indeed. Estimates vary as to the number of bees contained in a colony, but in a normal colony somewhere between 40,000 and 60,000 bees occupy a hive. Of these, there is one Queen only, a few hundred drones, and the remainder are worker bees. (*Fig. 13.*)

Taking these bees in order, we come first to the Queen.

The Queen Bee

A great deal of mystery surrounds the Queen, and

| Drone | Queen | Worker |

FIG. 13

enquiring strangers to the craft of apiculture invariably are anxious to see the Queen bee. She is the only genuine female in the hive, and is controlled by the workers and does not in any way rule the hive. She is born from a special cell known as the Queen cell. (*Fig. 14.*) This cell, the size and shape of a medium acorn, hangs downwards in the hive. The cell is formed by the bees of a mixture of beeswax and pollen, and is made semi-porous to enable the Queen to have air before hatching out. After the cell is formed, the workers usually themselves place in the cell an egg which has been laid by the Queen in a worker cell. Dr. Caird recorded having seen a Queen lay an egg in a Queen cell (*Bee World* 16, p. 69). It is quite understand-able that she herself will hesitate to lay the egg for a new-comer who will ultimately take her place in the hive. The workers immediately place in the cell along with the egg a white creamy substance which is called by bee-keepers, " royal jelly." This " Royal Jelly " has been found to have invigorating and tonic qualities. It is now produced by some pharmaceutical chemists in a palatable medicinal form, particularly in France. When the egg hatches on the third day, the grub which emerges from the egg feeds on this special food for a period of five days. It then spends a day spinning a cocoon, after which it rests for a further two days.

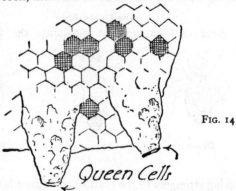

FIG. 14

Queen Cells

On the next day, the eleventh, it is transformed from a larvae into a nymph, and after three days as a nymph it emerges as a virgin Queen bee. The emergence of a Queen is a matter over which she has little control. The control is exercised by the workers, who only permit the Queen to emerge either to replace an old Queen who has passed her usefulness as an egg layer or to replace a Queen who has left with a swarm. It therefore takes 14 days to rear a Queen from the egg to the Queen herself. After leaving her cell, from which she is released by the worker bees, who lift the cap from the cell, she walks about the hive without any ceremony. After a few days, she goes out of the hive on the first available fine day on her first flight. The length of time she remains in the hive depends to a very great extent on weather conditions. Fine weather is essential. The drones in the hive and neighbouring drones fly in front of the hive containing the virgin Queen, and she comes out on the landing board of the hive. Eventually she takes her first flight. Observers say that she hovers in front of the hive to establish her location and then flies upwards into the sky, followed by any drones which may then be flying. Dr. Butler believes that the Queen makes several orientation flights before making her mating flight.

Coition between drone and Queen takes place in mid-air, and results in the immediate death of the drone, for his gential organs are torn out by the queen. She then returns to the hive, still having the drone's organs attached. She is welcomed by the workers, who immediately commence to feed the queen lavishly and remove the drone's organs. The author has seen a virgin queen leave the hive, circle round many times, and fifteen minutes later return with the drone's organs attached. In a few days the Queen commences to lay eggs, somewhat spasmodically at first, until later she lays at a very great rate. It is generally accepted that in her prime, that is in her second year, she lays at the rate of 2,000 to 3,000 eggs a day. The Queen is a handsome insect ; a long and graceful abdomen, but with short wings which fold neatly over her back. The laying power of the Queen is one of the marvels of the hive. For months on end she will lay at a great rate, and her useful life lasts roughly four years, but for honey production, experienced bee men do not keep the Queens for more than two years at the longest. It must, however, be borne in mind that during the coldest part of the Winter, the Queen either ceases to lay altogether or her laying is reduced to a very minimum. The bees, however, when the days lengthen towards the middle of February, begin to feed the Queen

B*

with more food, and she lays a little more frequently in the centre of the winter cluster. It is at this stage that a cake of candy over the hole in the quilt will provide the necessary supply of food to ensure the safe arrival of the young bees, which are so important at this stage to the future welfare of the hive. As the weather becomes warmer, so the Queen expands her laying until at the end of April she becomes the prolific layer she continues to be until mid-August. At the end of this period of the year her laying is curtailed by the bees, who feed her less generously, because at this season of the year the honey flow is ending.

The reader is referred to the chapter on swarming for further information regarding this most interesting creature.

The Drone

This is the male bee. The control of his life is in the hands of the worker bee. His birth is restricted in numbers by the worker. He is principally fed by the workers. He remains indoors during any but the best of weather, and when his functional services are no longer required, and when the honey flow draws to a close he is killed by the workers.

In the early Spring the workers gather in nectar from the Spring flowers, and the urge for colonisation enters into the hive. After ensuring stores for the future occupants of the hive, the workers then commence to build drone cells. Many of these may already exist in the combs in the brood chambers from the previous year. If they do not, the workers will break down some worker comb and rebuild it as drone comb. This conversion usually takes place in the worker comb, which is not of good quality, namely that at the bottom of the comb. These cells are usually made in the lower half of the comb, and are easily distinguished by their size, being larger than the worker cells. The Queen is then induced to lay eggs in these cells. The Queen lays an egg precisely similar in appearance to any other egg she lays, but the result is a male. How it is done has never been satisfactorily explained. The author in 1944 had the privilege of discussing this matter with a research student at the Hebrew University, Jerusalem. Her theory is that all the eggs laid by a queen are identical but that the sex of the resultant bee is determined by the quantity of Vitamin E which is supplied by the bees themselves in the food which they feed to the grub.

If her theory is correct—and the author feels sure it is from experiments which he has carried out both before and since meeting this student in Jerusalem—then one of the mysteries of the hive will at last have been solved.

A theory which has often been expounded is that the queen lays the appropriate egg for the cell. It is clear that there is more room in the drone cell in which the Queen can lay, and it is supposed by some that the act of the Queen laying an egg in a worker cell in a somewhat restricted position brings the egg into contact with the seminal fluid of the drone, which she constantly retains in her abdomen. It is also true that the virgin or unmated Queen which lays eggs lays nothing but male eggs, and so it must be assumed that the drone's mating has the effect of enabling the Queen to lay female eggs in addition to male eggs. Incidentally a laying worker bee, of which more is said elsewhere, being unfertilised, lays none but male eggs.

The egg laid in the drone cell hatches out at the end of the third day, and the larvae or grub is fed for six days by the workers. Three days are spent by the larvae in spinning a cocoon. Four days are taken to transform the larvae to the nymph stage, and the drone remains in the nymph stage for seven days, and on the 24th day it hatches out and comes out of the cell by its own efforts. Drones are often helped out of their cells by the workers and some are of opinion that this is usually the case. The capping of the drone cell is very readily distinguished by its domed appearance. The crown of each cell is raised fully 1/16th of an inch above the cell walls. It will be seen that the drone emerges from its cell on the 24th or 25th day after the egg is first laid. The drone remains in the hive at least 5 days before emerging for its first flight. But he may not be fully mature until 14 days old. This should be borne in mind when queen rearing. The drone can easily be distinguished in its flight by the noise it makes. Drones have never been seen to alight on flowers to help themselves to nectar, for they have not the organic faculties for carrying back honey to the hive. Although they make this noise in flight, which is somewhat alarming to the beginner, the drones have no stings. There are two points of view concerning drones some think that the bee-keeper should endeavour to curtail the number of drones in the hive, for they are said to have two ill effects. One, they consume honey which would otherwise be stored, and the other, that their presence in large numbers encourages the workers to prepare to swarm. This is particularly important, for if large stores are to be worked for it is essential that swarms should not be allowed to leave the hive. The other view is that drones are natural to a colony of bees and assist by generating much needed warmth. On the matter of swarm prevention the reader is referred to the chapter on swarming.

The Worker Bee

The worker bee is one of the most interesting, instructive and useful of all insects. By sex it is an undeveloped female. The grounds for this assertion are, first, that the worker is born of the same variety of egg as a Queen. An egg from a worker cell or even a hatched out grub from a worker cell may be placed by the bee-keeper in a Queen cell and a perfectly normal Queen will emerge. Secondly, in the event of a stock being without a Queen for a length of time, the bees will feed up workers as they do a Queen and the workers will commence to lay, but, alas, only male eggs and all the offspring are drones.

It is supposed that these converted workers can only lay an egg every other day or so. Laying workers result from a shortage of brood and a consequent absence of work for the nurse bees and not necessarily by any concerted action by the bees themselves.

Workers are only produced from eggs laid by a Queen after she has been fertilised. It is a source of great satisfaction to the bee-keeper who has been Queen rearing when he sees a patch of eggs and grubs in worker cells, but more so when he sees that the cells are covered with the characteristic cappings of worker cells. These cappings are very slightly raised above the surrounding cell walls and are made of beeswax and pollen and are khaki- or coffee-coloured.

When the Queen is fertilised by the drone, the seminal fluid from the drone is stored by the Queen in a small sac in her abdomen, and when she lays eggs her abdomen is curved so that as the egg is about to pass out it comes into contact with this seminal fluid which enters a small hole in the egg, called the micropyle, and the egg then becomes fertile.

The eggs, when laid, hatch out on the third day into minute pearly white grubs, and are then fed by the bees for five days, after which they spend two days spinning a cocoon, and remain at rest for a further two days. The transformation from larvae to nymph takes a further day. The last stage is a seven-day period in the nymph state until, on the 21st day after egg laying, the young worker bee hatches out. It emerges from the cell by its own effort, and these young bees can often be seen biting their way out through the cappings. They appear fluffy and young. The older bees appear sleek and glossy in comparison with the young ones. For the first day or so young bees spend their time feeding and becoming accustomed to their surroundings. They then enter upon their duties as nurse bees. They feed the grubs with " pap," which is a mixture of honey, pollen and water. Young bees are believed to feed the older grubs first and cannot feed the younger ones until the pollen they eat has developed their brood food glands. They feed also the Queen and help the foraging bees unload their burdens until, about the ninth or tenth day after birth, they emerge for a " sunning " on the landing board of the hive. At the height of the day, young bees can be seen on the landing board running about in circles and then " taking off " for their first flight, which consists in hovering in front of the hive. When a number of young bees do this it often stimulates the preliminary stages of emergence of a swarm. Suddenly the " party " ceases, and the bees return to their domestic duties.

Between 14 and 21 days after birth, the worker bee commences to work out of doors, but they commence as early as the sixth day if the foraging bees have been lost by the hive being moved. This work consists of water carrying, pollen and propolis gathering, and nectar gathering. These tasks are all tasks of bringing in to the hive. But there are important tasks of carrying from the hive, namely, the scavenging duties. These include the cleaning of the hive, the removal of any dead bees, and the removal of the faeces of the Queen, drones and young bees. Bees are the cleanest of insects. The hive is always sweet smelling except when diseased. They will never allow anything dead to remain in the hive if they have the power to remove it. If they are unable to remove anything from the hive they cover it with a coating of propolis, which is a sweet-smelling bee glue obtained from the sticky buds of trees. Mice and snails have been known to enter hives and die there and have been found to have been completely covered and sealed by the bees with propolis. As an example of their cleanliness, let the bee-keeper uncap some sealed drone cells. Hardly before he has completed replacing the covers of the hive the bees can be seen dragging the white corpses out on the landing board. The author has seen a worker bee get on top of the corpse and wrap its legs round it and with great effort take off with a burden much heavier than itself and fly upwards and away from the hive, and when about twenty yards away drop its burden and return to the hive. Again, one of the duties of the worker is to clean out and polish the interior of the cells in readiness for new eggs from the Queen.

The collecting duties of the bees are the principal duties of practical interest to the bee-keeper, particularly the collecting of nectar.

Pollen

Pollen collected by the bees is obtained from most flowers and catkins. It is the fertilising dust of flowers, it is usually found in large quantities on the summits of the anthers. The flowers depend in many cases for their fertilisation on the visit of pollinating insects. Nature provides encouragement to the bees by giving the flowers bright and gay colours, attractive perfumes and, not the least of attractions, nectar. When gathering pollen bees will pass from flower to flower of the same varieties. They will pass over attractive looking flowers to find flowers of the same variety as those they have just visited. This appears to be Nature's way of ensuring the continuance of existing species of flowers instead of producing innumerable hybrids. The pollination of flowers is now regarded by agriculturalists as the primary duty of the honey bee. Fruit growers are recommended by experts to have at least one hive of bees to the acre of fruit trees. The complete pollination by honey bees not only produces more fruit but it produces fruit of correct shape. Lob-sided apples, for instance, are almost unknown in an apple orchard correctly populated with bees. The Ministry of Agriculture, realising the value of the honey bee, recommends fruit growers to avoid the use of arsenical sprays when the trees are in flower.

The bees gather the pollen in their baskets, which are technically called "corbiculae." The bees pack the pollen in these baskets, which are on their hind legs. They pack in each hind leg a quantity of pollen about the size of a mustard seed. The pollen varies in colour according to the flower visited. For example, dandelion produces golden yellow pollen, poppy black, willow herb saxe blue, apple pale green, mignonette carrot colour, and so on. One of the interesting pastimes of the bee-keeper is to watch the bees at their work and watch what flowers are being visited and to discover whether nectar or pollen is being carried and to study the colour of the pollen. Although the bees are so particular in selecting the same flowers for collecting the same variety of pollen, yet, when they return to the hive, they pack away the pollen into cells regardless of colour. When a pollen cell is opened a large variety of colours can be seen in the same cell.

Pollen is called bee bread, and it is used by the nurse bees to feed the grubs which hatch out of the eggs. They masticate the pollen along with honey and produce a creamy liquid. Pollen provides the protein which provides the bodies of the bees with replacement of tissue.

The bees usually store pollen in the outer combs of the brood chamber, usually the one next to the outermost comb.

It is important when creating new stocks, as described elsewhere, that stores containing pollen should be provided to ensure that the new stocks are not handicapped for lack of essential foods.

Propolis

The word propolis means "before the city," or an outwork in defence. Before the introduction of the modern hive, bees were accustomed to close up the entrance of the straw hives with this substance, leaving apertures through which they left and entered the hive. This was done by them to prevent the intrusion of their enemies, in particular, according to Huber, to exclude the " Sphinx Atropos," commonly known as the death's head moth.

Propolis is a gum which is collected from the buds of trees and is used by the bees to stop up all cracks and loose parts in the hive. It is an extremely pleasant smelling gum, and it has a similar smell to vanilla. Propolis may become a nuisance, particularly when the bees use large quantities of it to seal the tops of movable frames to the brood chamber. This nuisance can be overcome if the beekeeper will use vaseline on the underside of the lugs of the frames and on the metal ends on the lugs. The bees collect this material and pack it in their pollen bags. Drones do not possess these corbiculae, and therefore are unable to collect either pollen or propolis.

The trees principally visited by the bees for this substance are the poplar, pine and horse chestnut. Bees have been found to collect bitumen from roadways and use this for the same purpose as the natural product.

Hives containing large quantities of this material are usually those placed under large trees. The motion of the roots of the trees in high winds causes the bees to secure the interior of the hive, and particularly the movable parts. Some species of bees, particularly Caucasians, use this material in large quantities.

Nectar

This substance is secreted from the the nectaries of flowers. This secretion is one of Nature's ways of attracting the pollenising insect. As the bee travels from flower to flower in search of this sweet liquid, its body comes into contact with the pollen. This is carried from flower to flower, reaching ultimately the ovaries of the visited flowers. Bees visiting flowers which are self-pollinating shake the pollen from the anthers on to the stigmas.

The bees suck the nectar from the flowers. It then

passes through the oesophagus into the honey sac, which is the storage organ of the honey bee. Before being collected by the bee, nectar is a dilute solution of cane and other sugars in water. It also contains aromatic substances, which provide the flavour and aroma of the finished honey. The sugar content varies considerably. The sugar content depends to a very great extent upon weather conditions, e.g., the humidity of the atmosphere rather than upon plant species. For instance, nectar from an apple-tree at 8 a.m. on any given day may have a concentration of 15 per cent. but at 3 p.m. on the same day may have a concentration of 30 per cent. This change is due to the evaporation of water leading to the concentration. Some flowers produce a 50 per cent. content, whilst others are recorded as producing less than 20 per cent. sugar content.

When the honey sac is full, the bee returns to the hive and regurgitates the nectar into the cells, where it accumulates. Sometimes the bee is in such a hurry to return to nectar collecting that it regurgitates the nectar to a nurse bee, who takes it up and then takes it and deposits it in the honey cell. When the nectar reaches the honey cell it has become a weak solution of grape sugar. The heat of the hive reduces the water content to such an extent that fermentation is impossible. When the water content has been reduced to the required standard, the bees will cap over the cell with a wax capping.

The secretion of nectar from the flowers varies generally according to the temperature of the day. It is usually supposed, for example, that it requires a temperature of 70° to produce the secretion of nectar in the common wild white clover.

The secretion is also affected by the rainfall. A long period of drought reduces the water content of the nectar and sometimes stops it completely.

The substantial secretion of nectar from a variety of blooms at one time is known as a honey flow. This occurs with spring blossoms, fruit blossoms, clover, lime blossoms and heather. It is most important that the bee-keeper should get his stock up to strength for each of these honey flows. They will not all occur in the same district, and the beginner should investigate the flora in his district so that he may prepare his stocks accordingly.

It is desirable, where the bee-keeper desires to separate his various grades of honey, that he should watch for the honey flows of the various flowers. For instance, when the apple and hawthorn are in bloom together, the bee-keeper should extract the sealed honey produced. This honey is considered by some to be the most delicious of all English honey.

The bees gather considerable quantities of honey from the sycamore and gooseberry, which are usually in bloom together. This honey is green, but is not considered of very high quality. The clover bloom provides a very fine quality honey which is very pale and produces a very attractive and finely granulated white honey. It is the most popular of English honeys.

The lime tree produces very large quantities of honey, but this crop is very fickle, starting about the first or second week in July and terminating at the end of July. Those who rely on this crop are often frustrated by a wet July, but when the weather is favourable surpluses of upwards of 75 lb. per hive from this source alone are not uncommon. The honey is pale green, and has a faint flavour of peppermint. It is a very pleasant honey indeed.

The main flow in moorland country is from the heather. This occurs in August, and produces a dark, thick and very popular honey. The demand for this honey is very great, and commands a price approximately 50 per cent. over other honey.

Other flora produce large quantities. Mustard and

charlock produce a pale honey which granulates quickly. The willow herb, which grows profusely where woodlands have been cleared, produces a very pale honey as clear as water. Lucerne, field beans, catmint (nepeta) and Michaelmas daisies are flowers which produce large quantities of honey.

Water

Bees gather large quantities of water for use in the hive, particularly to manufacture food with which to feed the young grubs. This at one time was, and still is, by some bee-keepers wrongly called " chyle " food. This term was originally applied to the brood food when it was believed that this was regurgitated from the " ventriculus " or " chyle stomach " of the worker. This is now known to be incorrect. It is known that one of the three pairs of salivary glands—the brood food glands—produce the brood food and that regurgitation from the ventriculus is physically impossible. They also collect this water for their own consumption, and for the consumption of the nurse bees which have not yet left the hive. Some experts state the quantity to be as high as a quart per day at the height of the season. However much it may be, it is essential that the bee-keeper should see that there is ample water available near to the hive. A water fountain can be made with a jam jar and a piece of grooved wood. This jar should be replenished daily, and kept at least 25 feet away from the hive, otherwise the bees will go to a much greater distance to obtain supplies.

The flying time spent in collecting water from a distance is of supreme importance. If this can be reduced, more is the time available for nectar gathering. Experiments have been carried out with great success by putting a rapid feeder over the hole in the quilt in the hive and keeping this replenished. It is quite surprising the amount of water which the bees will consume. It is recommended that one teaspoonful of salt should be added to each gallon of water. This is a mineral which the bees seem to appreciate.

Honeydew

In hot dry weather, bees collect this substance from the leaves of trees. It is only in hot dry weather that the bees are attracted by it. In normal summer weather, the bees will collect nectar from flowers in preference to it, but if there is a prolonged dry period, then the flow of nectar decreases and the bees then resort to honeydew.

In this dry weather the aphides which live on the underside of the leaves of certain trees, such as the sycamore, oak and plum, suck out the sap from the leaves and secrete a sweet sticky substance, and this is what the bees collect. This sticky liquid often falls to the ground and in dry weather can be seen in spots on pavements. The lower leaves of trees are often made shiny with drops of it which have dried on their upper surfaces. Motorists purposely avoid leaving their cars under such trees on account of the sticky drops which fall.

It is a rank-tasting substance, dark in appearance and, strangely enough, is stored by bees in cells separately from the honey. It can be detected in the comb by holding it up to the light. Before the combs are extracted, as described later in the chapter on The Honey Harvest, each cell which appears black, on being held up to the light, can be uncapped and swilled out with a fountain pen filler, so that the rest of the honey is not spoilt by this substance.

CHAPTER V

EQUIPMENT

IN order to carry out effectively simple bee-keeping, the following equipment is necessary.

(1) Hive

Different types of hives have been described elsewhere, but the beginner is advised to purchase or make the National or the W.B.C. hive. Whichever he buys or makes, he must continue with the same standards in order that his equipment can be interchangeable. This cannot be too strongly emphasised. An apiary having varieties of equipment cannot be efficient, and inefficiency in this way can only be overcome by expense. If the bee-keeper wishes to pay his way he must standardise.

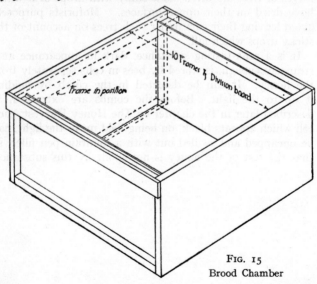

FIG. 15
Brood Chamber

(2) Brood-chamber

For the W.B.C. hive this must be of the B.B.K.A. standard to hold 10 standard brood-frames and one division board, the standard brood-frames having 1½ inch metal ends. (*Fig.* 15). For the National hive, 11 brood-frames are necessary.

(3) Brood-frames

The standard B.B.K.A. brood-frames should be purchased to complete the brood chamber equipment. Even though the bee-keeper purchases a stock on brood-frames he should have in reserve always 10 spare brood-frames, in case he picks up a stray swarm or forms a nucleus of his own. (See *Fig.* 6.)

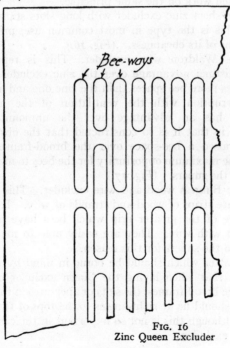

Bee-ways

FIG. 16
Zinc Queen Excluder

(4) Wax foundation for brood-frames

This should be purchased only according to requirements, as it deteriorates on keeping in that it loses both its colour, aroma and softness. Bees take much more readily to new wax foundation. However, the beginner should have sufficient sheets of foundation to meet any emergency. Nothing is more tantalising than to lose a swarm through lack of a little additional equipment. Eleven sheets of medium foundation weigh 1½ lbs.

(5) Queen Excluder

This is essential when extracted honey is being worked for. There are three recognised types of Queen excluder, all of which work on the same principle.

(*a*) The sheet zinc excluder with long slots stamped out of it. This is the type in most common use, principally on account of its cheapness. (*Fig. 16.*)

(*b*) The Waldron wire excluder. This is reputed to have a distinct advantage over the zinc excluder in that it provides more bee spaces than the zinc one and allows of less interruption with the ventilation of the hive. It certainly has an advantage over the unmounted zinc excluder, in that it is so constructed that the obstructing wire is raised a bee-space over the brood-frames. This affords the maximum opportunity for the bees to make their way into the supers. (*Fig. 17.*)

(*c*) The Burgess wood and wire excluder. This consists of alternate strips of wood and strands of wire. It has the advantage of the greater grip which bees have on wood compared with wire. They are easily able to make their way up to the supers for this reason.

One matter which should be borne in mind by the bee-keeper is that when laying the Queen excluder over the tops of the brood-frames, the slots or apertures in the Queen excluder should lie at right-angles to the tops of the brood-frames, although this is not so important in the case of the Waldron excluder.

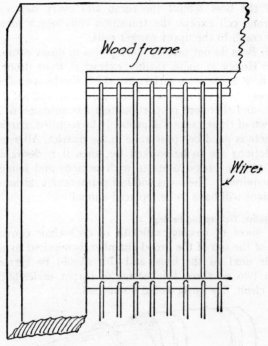

FIG. 17
Waldron Queen Excluder

(6) Shallow frame rack

This rack consists of 10 extracting or shallow storage combs, in the case of the National hive 11 combs. These should be fitted with drone foundation for three very good reasons.

(a) Bees in the wild store most of their surplus in drone comb. The author, as an experiment, recently used a brood-comb in the upper chamber when "demareeing." The lower half of the brood-comb had many drone cells in it. The lower half of the comb was cut away and during the honey flow the bees rebuilt the comb and every new cell was a drone cell except the transition cells which joined the new comb to the upper worker cells.

(b) Bees do not usually store pollen in drone comb.

(c) Honey is more readily extracted from drone comb as there is less wax surface to which the honey adheres in proportion to the amount of honey as compared with worker cells, and therefore proportionately less adhesion of honey.

Two of these racks will probably be required, particularly if there is good bee pasturage in the district. Alternatively, if sections are to be worked for, then it is desirable that two section racks complete with sections and foundations be acquired. The racks will last permanently, although the sections will have to be replaced annually.

(7) Quilts, Ticking and Felting

A sheet of ticking, sailcloth or deck-chair canvas, the size of the top of the brood-chamber, is required to act as a cover next to the bees, and this should be surmounted with two or three thicknesses of carpet under-felting or two clean, dry sacks would do instead.

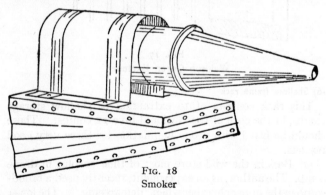

Fig. 18
Smoker

(8) Smoker

This is a necessity, especially to the beginner. It gives

one a feeling of confidence when approaching a hive to feel that one is armed with an instrument which will subdue bees. (*Fig. 18.*) In time, however, the bee-keeper will discard this instrument, particularly during the honey flow, as the effect of smoke on bees is to make them gorge themselves with honey. Occasionally the bees acquired do require the smoker with which to subdue them.

(9) Carbolic Cloths

Two of these cloths, the size being slightly larger than that of the brood chamber top, are very useful. They should be soaked in a 10 per cent. solution of Jeyes' fluid or carbolic acid, and kept moist in a tin. The reason for and the use of these cloths, which are commonly called stink cloths, is described in the chapter on Subduing of Bees.

(10) Feeder

There are many types of feeders, and they are used for a variety of purposes, but the one of most service to the bee-keeper is the rapid feeder. Although this is comparatively inexpensive, it can be dispensed with to curtail expenditure. A very convenient makeshift alternative is a piece of perforated zinc placed over the feed hole in the quilt and a 2 lb. jam jar, the mouth of which is covered with muslin. (See *Fig. 30.*)

(11) Perforated Zinc

A piece of this material 6 inches square, which is useful for wintering and for the makeshift feeder, should be purchased for each hive.

(12) Veil

This is a necessity. There are two principal types :—

(*a*) Fine black cotton netting which can be purchased from any draper. A yard 36 inches wide is sufficient.

(*b*) The wire veil. This is much more durable and

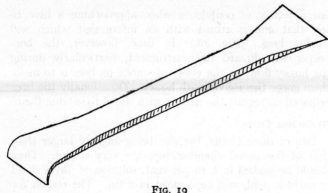

FIG. 19
The Hive Tool

cheaper in the long run. Further, it gives better protection.

(13) Hive Tool

A very handy thing to have, but it is not a necessity·
A blunt, old wide chisel will serve equally as well. This is
used for levering open various parts of the hive which have
been glued together with propolis. (*Fig. 19.*)

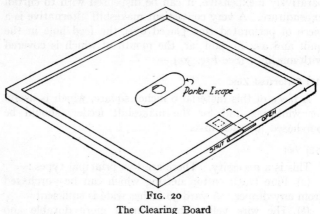

FIG. 20
The Clearing Board

(14) Clearing Board

This board, sometimes called a bee escape board, is desirable but not essential. The super can be cleared with stink rags, as described elsewhere in the chapter on Subduing of Bees. If, however, the bee-keeper intends to keep more than one stock, he should purchase or make one of these. (*Fig. 20.*) These are often called " Porter Board."

Diagram 21 shows the detail of the Porter Bee Escape,

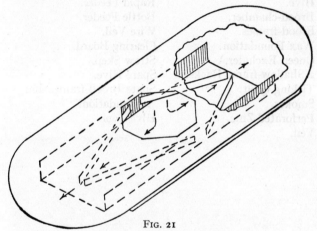

Fig. 21

Details of the Porter Escape

which is set in the middle of the clearing board.

There are many other items of equipment which may be purchased, but they are not necessities. If the bee-keeper will join the local association, he will almost invariably have the advantage of hiring for a nominal sum the Association's extractor and ripener. And here a word to those who borrow this equipment. Do please scald out this equipment before returning. Take the extractor to pieces, scald each part, and wipe dry. Scald out and wipe dry the barrel and reassemble, and remember that it will last much longer if it is cared for. So many who borrow the extractor never dream of taking it to pieces—a simple operation—to clean it properly. Honey has certain natural acids in it, and these, in a very short time, will penetrate the tinned surface of these implements and rust will ensue.

To summarise, here is a table showing what equipment is essential and what desirable.

Essential Equipment.	*Desirable, but not Essential.*
Hive.	Rapid Feeder.
Brood-chamber.	Bottle Feeder.
Brood-frames.	Wire Veil.
Wax Foundation.	Clearing Board.
Queen Excluder.	Straw Skep.
2 Shallow-frame Racks.	Spare Hive.
Ticking Quilt.	Spare Brood-frames and
Smoker.	Foundation.
Perforated Zinc.	Hive Tool.
Veil.	

bee-keeper. He can handicap them by their looking for trouble.

example, that all the combs upon which the bees are working are two inch drone-cells probably. When the bees have copied upon these drone cells a larger sized comb about a...

CHAPTER VI

SWARMING AND SWARM PREVENTION

THIS is a perfectly natural propensity of all stocks of bees. It is the only natural method of perpetuating this species of insect. The bee-keeper, however, does not desire the bees to spend their energy in colonising—for that is what swarming is—but rather to direct it into the storage of surplus honey.

It has been found by long experience that the bee-keeper can exercise considerable control over the swarming instinct, apart from the selection of supposed "non-swarming" varieties of bees.

The first signs of swarming can be detected with fair accuracy. In the early Spring, when the first blossoms provide the bees with nectar, breeding of worker bees proceeds with increasing speed, and storage of surplus nectar commences. It is then that the bee-keeper takes the first step in swarm prevention. This step is to provide the bees with more storage room. This action of the bee-keeper may delay the swarming impulse, for the bees now have more room in which to operate. But the natural impulse cannot be frustrated for long. The bees realise, presumably from instinct, that when they swarm they must leave behind males who will fertilise the queen whom they will never see, but who will be born when they have left the hive. They again, by instinct, know that the males take longer to be created than Queens. They therefore show the next sign of swarming by building a number of drone cells, and at this stage they will even break down a number of worker cells and rebuild drone cells in their places. This diagram shows worker cells on the left and drone cells on the right, with transition cells between the two. (*Fig. 8.*) This should be watched by the bee-keeper. He can handicap them in their intentions by ensuring that all the combs upon which the bees winter are as free from drone cells as is possible. When the bees have capped over these drone cells, the bee-keeper should uncap them so as to expose the sealed larvae or nymphs. The bees will then not tolerate these exposed corpses in the hive, and within a very few moments can be seen hauling them out of the hive, flying away with them, and dropping them in the distance. This is regarded by some as a bad practice and not tending to prevent swarming but the author finds in practice that it is effective. This action defers the progress of the creation of drones for several days. When destroying the drones in this way, the bee-keeper should search very carefully indeed for queen cells, and if any should be found, they should be cut away from the combs in which they are built. At least every 10 days the hive should be inspected for signs of queen cells, all of which should be ruthlessly removed.

Weather conditions may also help the bee-keeper. Rain will also prevent bees swarming, but here the bee-keeper should be warned. When bees are kept indoors in the early part of the year, they often rapidly build queen cells in readiness for the fine weather to come. If, however, the weather remains bad for any length of time, the bees may give up the idea of swarming and destroy the queen cells and their occupants.

If the bee-keeper is unable to prevent a swarm and the weather is favourable, the swarm will emerge. Those who have seen a swarm emerge from a hive will agree that it is one of the most astonishing of phenomena in nature, and this is what is seen. A number of bees will hover in front of the hive facing the hive. Among these will be seen a large number of drones creating their boisterous noise. This noise is said by some to stimulate the impulse of swarming. Those who have observed the interior of an observation hive say that the queen ceases to lay and can be seen hurrying about the hive from comb to comb. Eventually she comes out on to the landing board. Meanwhile thousands upon thousands of bees have poured out of the hive. The appearance of the mouth of the hive resembles boiling treacle. There seems no end to the exodus. They crowd the landing board and the front of the hive and take off into the air, swirling round and creating an alarming noise. Gradually the centre of the swirling crowd of bees moves away from the hive and becomes more compact. The queen has left the landing board and taken to the air for at least the second time in her life, the first being her mating flight. Whilst this is stated here the writer cannot with certainty say that she never leaves except on these two occasions. That appears to be the general opinion of many. On the contrary it may be that she may take a trip on her own. Since the publication of the first edition the author has received many letters which satisfy him that Queens do in fact make excursions after the mating flight. Hives are often found to be queenless without any apparent reason. It may be that the Queen has either been destroyed by the bees or else has taken an occasional flight and not returned having either lost her direction or having been destroyed by birds.

Suddenly there will be noticed on the branch of a shrub or tree a small knot of bees which grows very rapidly as the flying bees settle. The cluster varies in size according naturally to the number of bees which leave the hive. It is, however, very common for swarms to be about the size of a Rugby football, and to weigh five to six pounds. There are between 3,500 and 4,000 bees to the pound, and therefore a swarm of five pounds weight approximates 20,000 bees. Bees weigh heavier when swarming as they fill themselves with honey before leaving the hive. The Queen is in this mass. If she leaves and flies away, the bees will follow her.

c

This swarm may hang on the tree for as long as 48 hours, but this depends principally upon the weather. The bee-keeper is advised to take the swarm as early as possible.

A day or two before the swarm emerges the stock has sent out scouts to find a suitable new home. This they find, investigate it, and prepare it for the swarm. They then return to the hive and when the swarm emerges and

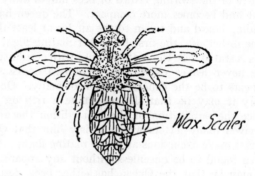

FIG. 22
Worker Bee with Wax Scales

has settled down from the excitement of swarming, lead the swarm to its new home. It is therefore desirable to skep the swarm as early as is possible.

Also whilst the bees are clustering in this mass they become hot and commence to sweat the wax through the eight glands on their abdomens in readiness for the new home they are going to furnish with honeycomb. *Fig. 22* shows an enlarged drawing of a worker bee with the wax scales on the underside of the abdomen.

Here follows a description of the method of taking a swarm.

How to take a Swarm

At first the prospect of capturing 20,000 bees appears formidable, and a little courage is needed by the beginner. However, properly protected, the bee-keeper will come to no harm. The veil should be worn, and gloves also, particularly if the bees are in a difficult situation. A man should either put his socks over his trouser leg bottoms or wear cycle clips or gum boots. A woman should wear slacks or jodhpurs, and her arms should be covered. A golf jacket is admirable. Bees do not like rough woollen clothes, and will often attack the bee-keeper who wears them. They also appear to dislike navy blue clothing. This protection should be used by the beginner to prevent him being badly stung. Later he will get over the fear of being stung and should discard all except the veil.

It is not always necessary to use a veil when taking a swarm but the beginner should do so.

Before the bees come out of the hive in the swarm they consume considerable quantities of honey. This makes the bees content and less likely to sting the bee-keeper. There are some writers who say that they consume enough honey to last them three days, but it is quite impossible for a bee when active to consume sufficient honey to last it for three days.

To take a swarm, the equipment needed is a white cloth 4 to 5 feet square, a straw skep [see *Fig. 2*] or an open wooden box about 15 inches square and one foot deep.

Spread the white cloth on the ground in the shade near to the swarm; take the skep or box, and hold it mouth upwards immediately under the swarm. Take hold of the branch on which the swarm is hanging and give it one or two firm shakes so that the swarm, or the greater part of it, falls into the skep or box. The skep or box should then be carried into the shade, and placed mouth downwards in the centre of the sheet. Take two stones about two inches thick or two pieces of wood and place them under one side of the skep or box so as to allow the bees to come in or out. If the Queen is in the skep or box then the bees will remain with her there. If not, the bees will come out and fly back to where she may be, or if the Queen is lost they will return to the hive. Stand by for a few minutes, and it will be noticed that the remaining bees on the tree will join the swarm in the skep or box.

The skep should be sheltered from the sun and left until the late afternoon. If the bee-keeper wishes to allow this swarm to start a new stock, he will prepare a hive in readiness to hive it, or he may wish to return the swarm to the hive. There is always a risk that the swarm may leave the hive before evening and many bee-keepers hive their swarms straight away.

Hiving the Swarm

The following is the method of hiving the swarm. First prepare a hive in the following way. Have a brood-chamber with 10 brood-frames with foundation fixed—or for preference ten drawn out brood-combs and a division board. Place this on a hive floor board and raise the front of the brood-chamber about one inch by means of small wedges of wood. Cover the brood-frames with a quilt having a feed hole in it. In front of all this and sloping up to the hive place a large board.

Go to the swarm at dusk. Remove the stones or small blocks of wood, take hold of all four corners of the white cloth and tie them over the top of the skep or box and carry the whole and stand it on the board sloping up to the brood-chamber. Untie the white sheet and so arrange it that one edge almost reaches the entrance of the brood-chamber. Level out the sheet so that it lies flat on the board. Take hold of the skep or box firmly with both hands and raise it a little over a foot from the white cloth and with one sharp jerk throw the contents downwards on to the cloth, retaining hold of the skep or box. A firm throw will invariably dislodge the whole mass of bees.

This seething mass will then turn their heads uphill, as it is the natural tendency of bees to crawl uphill, and some will approach the entrance of the hive. These then come out of the entrance, face the hive, turn up their tails and fan their wings very quickly, exposing a white speck at the point of the abdomen. This white speck is the scent gland. The fanning bees at the entrance to the hive blow the scent towards the mass of bees, thus directing them to their new home. Then the stampede starts. Look for the Queen and you will see her enter, Once she has entered, all the others will follow.

When all the bees have entered the brood-chamber the wooden wedges should be removed and the outer walls and roof of the hive should then be placed round and over the brood-chamber. The feeder, for preference a rapid feeder, which should have been placed over the feed hole in the quilt and be filled with warm syrup of a strength of 1 lb. of sugar to a pint of water before the swarm is hived. This is most important for two reasons. First, the bees need food to enable them to produce wax to use in the drawing out of the comb. The sooner the comb is drawn out the more quickly will the Queen commence laying. Whatever is used by the way of sugar at this time will be amply repaid. It is truly an example of " casting bread upon the waters." The supply of syrup should be kept up for a few days until it is observed that the bees are well established, then take away the feeder and place on top the queen excluder and a super, as the bees which have swarmed store very rapidly after the foundation is drawn out.

The second reason for feeding is to secure that the bees remain in their hive and do not abscond. Many a beginner has discovered that the bees have flown, and upon enquiry from an experienced hand has learnt once and for all of his reason for disappointment.

Hunger Swarms

Occasionally a bee-keeper is called to remove a swarm of bees of unknown origin.

The bees may appear vicious and restless. If the time is very early spring or late in the season, a hunger swarm should be suspected. The method of dealing with this is to take a skep or box and smear it inside with honey. Shake the swarm into the skep, and place a feeder immediately on top of the skep and leave it overnight. The swarm should only be hived when it has become more docile. To avoid any wasted effort, the bees can quite properly be left to winter in the skep provided they are well fed. In the early spring the skep should be placed on top of a brood-chamber filled with brood-frames fitted with foundation. The bees will gradually work their way down into the brood-frames, filling the skep with honey in natural honeycomb. [See Plate I.]

How to prevent Swarming

(1) Selection of Stock

The swarming instinct is stronger in some strains of bees than in others. Therefore the first step towards swarm prevention is to acquire a stock from a breeder whose bees are not prone to swarming. Many dealers advertise bees as " non-swarming." This is a misnomer, for all strains of bees will swarm if uncontrolled.

It has been stated earlier that bees show signs of swarming in advance of that event.

(2) Destruction of drone cells

The bee-keeper should examine the hive early in the spring to see whether the combs have many drone cells.

If any combs have a large number of these, they should
be removed and new brood frames fitted with worker
foundation. In any event three combs out of ten should
be replaced each year, so those having most drone cells
should be removed. If the Queen has laid eggs in drone
cells it is almost certain that the bees intend that there
shall be plenty of prospective mates for the Queens which
the workers will rear in queen cells. The bee-keeper
should uncap any drone cells which are sealed over. The
cells can readily be identified as they are larger than the
worker cells and have a dome shaped capping. An
ordinary pocket knife is all that is necessary. Within a few
moments the bees will have removed the white corpses
of the drone nymphs from the hive. If all capped drone
cells are removed, it is almost certain that the bees will
start again to produce drones, but the process of swarming
will have been deferred for a week or two.

(3) Removal of Queen cells

Shortly after the Queen lays in drone cells the workers
will commence to build Queen cells. How long after can
approximately be determined as follows. A drone is
ready for function as a mate for a Queen when it is 14
days old. A Queen is ready for mating when it is
approximately 4 days old. Therefore the drone must be
born at least 10 days before the Queen. A Queen takes
approximately 15 days to mature from the time of
laying the egg. Therefore the Queen cell is charged
with an egg five days before the birth of the drone.
As it takes a day or so to construct the Queen cell,
it can be regarded as certain that if drone cells are
allowed to progress beyond 14 days the bees will start
to make Queen cells.

Again, the Queen cells, when seen, should be neatly
cut out with the pocket knife. They are easily seen.
They hang down from any place on the comb or even

from the wooden frame, but particularly do they hang from the bottom half of the comb. Frequently the bees cluster round them and hide them from the bee-keeper's view. The bees can be brushed away with a goose feather or blown on by the bee-keeper. The bees do not like human breath, and run away quickly, exposing the naked comb, and any Queen cells can readily be seen. All Queen cells are not shaped like a complete acorn. In the early state of construction they have the appearance of acorn cups. These should also be excised. Great care must be taken to ensure that all Queen cells are removed, for if one is left the stock will probably swarm just as if there were many.

The bee-keeper must inspect his hives for Queen cells at least every seven days.

(4) Provision of more room

The removal of drone and Queen cells only delays the swarming impulse, whereas if these steps are combined with other measures they become really effective. The first of these is the provision of more room. This enables the bees to spend their energies on building up their stores in the newly created accommodation.

The bees can be given more room in two distinctly different ways. The simpler method is to place a crate of shallow combs immediately above the brood-chamber, placing the Queen excluder in between the brood-chamber and the shallow combs. The bees will proceed to fill these with honey instead of filling the brood-frames which crowds out the bees. Some bee-keepers lift up the brood-chamber and place a crate of shallow combs of worker cells under the brood-chamber. This has its disadvantages in that the bees proceed to store honey in the brood-frames and to commence breeding in the shallow combs. It is doubtful whether the bee-keeper is ever able to extract this honey without it becoming contaminated with brood.

However, the bee-keeper who practises this latter method should only place the shallow crate below the brood-chamber where the Queen is a good layer, then he should also place a shallow crate over the brood-chamber. Very soon the queen will make full use of the laying accommodation, and if the season is good, will rapidly fill the supers.

(5) The Demaree system of swarm control

In the year 1892, a certain Mr. Demaree wrote to the *American Bee Journal* setting out his system of swarm control when the object of the bee-keeper was honey production. It is a system now widely practised in this country and its popularity is due to its simplicity and its general success. However, if carried out as suggested by the originator, it is a dirty method of honey production for reasons explained later. It is, however, an excellent method of producing natural stores in brood sized combs, useful for helping out weak stocks and providing winter feed.

This is the system. First go through each comb of the stock and remove all Queen cells in whatever state of development. Take out all combs with the exception of one having brood and place them in a clean prepared second brood-chamber, which is to be the upper storey. The brood-frame containing the Queen should be retained in the original brood-chamber. The number of brood-frames in the second chamber should be made up to ten in number and a division board fitted at one end. Likewise, the original brood-chamber should be filled up with brood-frames. More progress will, of course, be made if the added brood-frames have drawn out comb. Again, a division board should be fitted to this brood-chamber, and care should be taken to see that it is fitted at the same end as the one fitted in the second brood-chamber. Now place a Queen excluder over the first brood-chamber and over this place the second brood-chamber. In this way the Queen has a completely new brood-chamber in which to lay and continues laying at a rapid rate. Meanwhile, all the bees in the brood-combs above are hatching out rapidly and these young bees provide the Queen with a constant supply of nurse bees. All the cells in the upper storey will be vacated within twenty-one days of the operation being carried out. The hive will have an enormous population with a Queen laying rapidly. If there has been a honey flow most of the cells in the upper storey will have been filled with honey as soon as they are vacated by the emerging bees. If the honey flow still continues the combs in the upper storey should be extracted and the operation repeated. Dr. Butler has discovered what he considers to be an improvement on this system. He always places a shallow super complete with combs between the brood-chambers at the first operation.

The advantages of this system are, first, that the Queen is never cramped for laying room and therefore one of the predisposing causes of swarming is removed ; and secondly, that the system provides a large stock of bees for honey collection.

The disadvantages are first, that unless the operation is repeated after 21 days there is every possibility of the hive being so crowded that swarming is merely postponed and the swarm may leave at the height of the honey flow. Secondly, the honey produced is regarded by some as " dirty honey." It has been stored in cells where grubs have developed into nymphs and left their excreta behind and nymphs turned into bees in their turn having left their outer skins behind. They say it is not honey which a clean bee-keeper would eat himself and therefore he should not sell it to the public. This particular objection may perhaps be negatived when we are told by observers that during a good honey flow the field bees usually place the honey they have collected in the first available cell in the brood-chamber often on top of an egg or even a young larva. One very helpful letter which the author received after the publication of the first edition read as follows :—

" It is surely silly to call honey from brood-combs ' dirty.' The bees would be apt to be poisoned themselves (they would naturally be more liable to suffer than animals like ourselves not nearly related to them) if that were so. They do use old brood-combs for storage, in a wild state ; and natural selection would long ago have eliminated the habit if the honey was really ' dirty.' Do you eat sausages ? If so, is not that habit worse ? The vast bulk of commercial honey is produced in old brood-combs."

Thirdly, there is one danger which is not always explained about this system—and that is that unless every care is taken to prevent Queen cells maturing in the upper storey after the transfer to it of brood, a Queen may hatch out which cannot get through the excluder which may result in either a swarm emerging after all or the virgin Queen becoming a drone layer, as she is unable to get out of the hive to become fertilised. Therefore it is essential that after two or three days the bee-keeper should go through the upper storey and cut out any Queen cells which may have been formed in the interim. This is most important and should be repeated after a period of seven days.

There are those who think that if the Queen cells are cut out of the upper storey that the bees cannot place a young enough grub or egg in the upper storey. The writer's experience shows that they will fetch eggs from the lower storey even 10 days later.

This system is often used when the bee-keeper desires to increase his stocks, for in the upper storey the bees will often after the operation has been carried out make a large number of Queen cells. These can be made into nuclei which will be helped on rapidly by the large number of young bees which will shortly hatch out.

(6) Clipping the Queen's Wings

This is a method which is intended to prevent swarming by prohibiting the Queen from leaving the hive. There are two principal methods of carrying this out. First, the Queen should be located on a comb and whilst moving about should have one wing clipped by cutting with scissors half way along the wing. The second method is by picking up the Queen by the wings with the right hand then take hold of her with the thumb and first finger of the left hand by the thorax. Then one wing should be clipped halfway along. Both these methods have a grave element of risk and danger for damage to the Queen will probably have harmful results upon the progress of the stock.

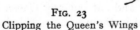

FIG. 23
Clipping the Queen's Wings

This method of swarm control does not prevent the swarming instinct for the swarm will emerge in just the same way as a natural swarm except that the Queen will not leave as she will be unable to fly. The bees then return to the hive when they find she is absent from the swarm. It may be however that the Queen has fallen from the landing board of the hive and unless careful watch is made she may be lost. However, if the bee-keeper sees the swarm emerge and finds the Queen on the landing board of the hive and returns the Queen and bees to the hive he should then take steps to overcome the swarming instinct by removing all Queen cells and giving the bees more room for honey storage.

(7) Artificial Swarms

The chief danger of swarming is that the swarm will be lost, and the chief disadvantage of swarming is that the stock is so reduced in strength that it is rendered virtually useless as a honey producer. If, therefore, the stock shows continued intentions of swarming it is desirable that the bee-keeper should form an artificial swarm. This has two main advantages, one that the bee-keeper's stocks are increased, the other that the swarm having been created artificially the urge to swarm dies.

This is how it is done. About the middle of a bright warm day when the bees are flying well, get a second empty hive and place it on the stand of the hive to be operated upon, removing that, the parent hive, to a distance of two or three yards. Then open up the parent hive, find the Queen and place the brood comb on which she is found into the brood-chamber of the new hive. Take from the parent hive also two other brood-combs containing stores of honey and pollen and place one at each side of the one containing the Queen. Then fill up the new brood-chamber with brood frames, having either drawn out comb or foundation. Similarly fill up the three spaces created in the parent stock with brood frames. Thus the new stock has a laying Queen and a number of young bees on the brood-combs along with the Queen. The bees which are out flying and collecting stores will return to the old stand and join the old Queen. Not a bee need be lost in the process. Now this is what happens to the parent stock. The bees finding that they have lost their Queen will proceed to make Queen cells if they have not already done so and will in time hatch out a Queen, but time may be saved by the introduction of a new Queen or by inserting a ripe Queen cell in the brood-chamber.

If this process is carried out in May or early June there is no reason why each stock should not build up sufficiently to take a useful part in the July honey flow and they will most certainly be ready for the August heather honey flow.

To assist the artificial swarm to make rapid progress they should be fed with a pint of warm stimulating syrup every other night for 10 days and the system of brood-spreading described in the chapter on spring stimulation should be carried out. If the bee-keeper is only able to provide foundation for the brood-frames instead of drawn out comb then a little more syrup should be given. This helps them to build the comb on the foundation more rapidly thus giving the Queen more room in which to lay.

If the bee-keeper desires still further to increase his stock he may repeat this procedure, using the parent hive again after a four weeks rest, being sure first of all that there is a reasonable quantity of brood and particularly newly laid eggs left behind, otherwise the parent stock will be rendered Queenless.

(8) Ventilation

One of the causes of swarming is congestion in the hive. Congestion causes heat and absence of air suitable for the bees to breathe. Therefore in hot weather it is desirable to see that the hive is kept as cool as possible and given extra air. The hive should normally be placed where it can be shaded from the mid-day sun, but in very hot weather the entrance to the hive should be shaded, but so as not to obstruct the entrance to the hive. The brood-chamber should be lifted on small blocks of wood about three-quarters of an inch thick and the roof of the hive lifted by putting two laths of wood on the top of the top lift and resting the roof on the laths. In this way there will be a through draught of air. As the weather cools the parts of the hive should be replaced otherwise the bees will remain indoors when they might be out collecting winter stores.

(9) The Snelgrove Method

This method of swarm control has proved itself to be the greatest boon to bee-keepers. It is an advance and elaboration upon the Demaree system, and those who practise it according to instructions are wholly satisfied with its results.

It would be unfair to Mr. Snelgrove to describe the system in detail in these pages and the reader is strongly advised to buy Mr. Snelgrove's book, " Swarm Control."

The exclusion of young bees from the brood chamber and returning them by a simple device is the principle on which Mr. Snelgrove works. His contention is that a super-abundance of young bees in the brood chamber is one of the principal factors in creating the swarming impulse.

His system works more readily with the National hive than with the W.B.C., although it can be worked with the latter.

The only additional equipment required is a board which is placed over the super but under the original brood chamber. This board has an area of wire cloth inserted which enables the top queenless stock to have the same smell as that underneath. This is important as it ensures the ready return of flying bees to the bottom stock from the top stock. By a simple system of opening and closing three pairs of entrances in the frame work of this special board foraging bees leave the top stock. Later the entrance is shut and the entrance into the lower stock opened, and so the foragers from the top stock augment the bottom stock which then becomes a powerful honey getter. The other two pairs of doors are similarly put into operation.

It is an almost foolproof system and admirably suited for out apiary stocks.

In time, naturally, unless prevented the top stock will hatch out a Queen which may be used for re-queening the parent stock.

These are the ways recognised which assist in swarm prevention.

Let Alone Bee-keeping

In the face of the current systems of swarm control there has grown up the cult of " Let Alone " bee-keeping. Once spring cleaning has taken place those who carry out this system of leaving matters alone in the brood chamber never interfere. They contend that constant interference disturbs the bees and that provided you give the bees plenty of room in which to breed and plenty of room to store, you will in the end obtain better results.

The author recently spoke to a lady who has always been a let alone bee-keeper even to the extent of her saying that she had never seen a queen bee in over 40 years' bee-keeping.

The author said ; " You must get a large number of swarms." " Yes " the lady replied, " but I get an awful lot of honey."

CHAPTER VII

HONEY PRODUCTION

HAVING, presumably, succeeded in saving the bees from leaving the parent hive, the bee-keeper will proceed to carry out his main object, namely, that of honey production.

It will be recollected that one of the principal ways of preventing swarms is to give more room. This is to enable the bees to store their honey in the room provided rather than in the brood-chamber. As the honey begins to accumulate in the super, naturally there is much less room for the bees. This must be remembered by the bee-keeper. As the honey becomes sealed, that is, covered with a white wax capping, the bees start to fill up the other cells. Usually the bees fill up the centre combs first. When these centre combs are almost filled they should be lifted and placed on the outsides and the remainder moved to the centre. In this way the bees are kept active with work always in front of them. When the first super is a little over half full, a second super should be provided. There is a right and wrong way of putting this on the hive. First have the second super complete with 10 shallow frames fitted with drone foundation or with combs already drawn out. The roof and outer lifts should be removed leaving the brood-chamber accessible. If possible the bees should not be smoked as this makes them consume honey. If, however, they are fractious, three puffs should be made with the smoker at the entrance to the hive. The existing partly filled super should then be slowly levered up with the greatest gentleness for rough handling disturbs the bees more than anything else. Before lifting the super, after loosening it, give it a slight twist in a circular movement so that when lifted it does not bring away with it any brood-frame which may have become attached by brace-comb. The super should then be lifted off and placed on top crossways on one of the lifts. A carbolic cloth should then be laid over the exposed Queen excluder which lies on top of the brood-frames. Next the empty super should be held immediately over the carbolic cloth. Care should be taken to ensure that the direction of the shallow frames should be the same as the brood-frames. The idea in placing the carbolic cloth on the top of the brood-frames is to drive the bees down to prevent them becoming restive and being crushed. Pull the carbolic cloth smartly away and gently lower the new super until it fits exactly on the brood-chamber top. Then replace the partly filled super, brushing away any bees which may be adhering to the bottom of the crate with a goose feather. Do this very gently, and brush the bees into the empty shallow frame crate. Close the hive down gently so as to avoid any further disturbance of the bees. The new crate is placed under the partly filled one in order to make the bees accustomed to the new crate when passing through to the old crate which they are in process of completing. An inspection a few days later will show that the bees are busy working on the new combs. They will in fact have commenced to draw out the wax foundation, or if combs already drawn out are supplied there will be signs of honey being deposited in the centre combs. When the clover is in full flower about the third week in June, honey will be stored more rapidly than before and the top super will be completed. If the season is good the top super will soon be filled. This should be removed, when about two-thirds of the cells are capped. The uncapped honey to all intents is ripe at this stage and a great saving in honey will be effected by relieving the bees from capping the remaining cells. The shallow frames should then be extracted and returned to the hive underneath the other crate of shallow frames next to the brood-chamber. When this is done the opportunity should be taken to place the unfilled combs in the centre of the other shallow frame crate.

Dr. Butler is of opinion that the effect of a carbolic cloth is that it subdues bees in exactly the same manner as smoke, that is, it makes them fill their honey stomachs with honey. But the writer's experience is that as soon as the carbolic cloth is placed on the brood-frames the bees run away from it without making any attempt to fill up with honey. On one occasion the writer had a powerful carbolic cloth on the hive and left it too long and it drove the whole stock out of the hive. Unfortunately the Queen was lost, so powerful cloths are no longer used by the writer.

It is convenient at this juncture to describe the effect of using wide metal ends on the lugs of the shallow frames.

These wide metal ends should be fitted when the foundation is drawn out almost to the fullest extent when the $1\frac{1}{2}$ inch metal ends are used. The wide ends are 2 inches wide and therefore they permit the comb being drawn out quarter of an inch on each face of the comb. The internal measurement of a shallow frame crate is usually 16 inches and therefore only eight shallow frames with two inch metal ends can be inserted instead of ten having the narrow metal ends. A shallow comb with narrow metal ends when completely full produces approximately $2\frac{3}{4}$ lbs. of honey, but a shallow comb having wide metal ends produces about $4\frac{1}{4}$ lbs. of honey. The comparisons are therefore 10 shallow frame combs at $2\frac{3}{4}$ lbs. each equals $27\frac{1}{2}$ lbs., and 8 shallow frame combs at $4\frac{1}{4}$ lbs. of honey each equals 36 lbs. Thus showing that the eight with wide metal ends hold more than ten with the narrow ends.

It is recognised that wax requires many times its weight of honey to produce. In consequence eight frames which have only a total of 16 faces to be capped as against 20 faces of the frames having narrow ends certainly afford a great saving of honey which would otherwise be used in producing wax.

There are those who argue that it is difficult to extract

honey from the combs that are drawn out so wide, but the answer to that is that if drone foundation is used for these combs the honey comes out very easily.

The two advantages of using wide metal ends are ; first that it is more economical in that only eight frames are needed instead of ten, a saving of 20 per cent. on shallow frames alone, and secondly, a greater amount of honey is produced because there is a saving in wax cappings.

The Honey Harvest

To the bee-keeper whose aim is honey production, this is the climax of his years' work. Assuming that as the bees have stored their honey in the combs provided, and the bee-keeper has continued to provide more storage space as and when inspections of the hive have justified it, the time will come when he should remove the surplus stores. As stated earlier, some bee-keepers believe in the removal of surplus as and when it is completely sealed over and replacing the frames taken out by frames fitted with foundation or with combs already drawn out. The principal argument against this is that it is better to allow the honey to remain in the hive and mature in the comb. There does not appear in any of the books which have been written on this subject any reasoned argument as to why sealed honey should acquire any more virtue after it has been sealed. There are however several very good arguments why it should be taken from hives piecemeal. Here are the principal reasons. The capital outlay is much less for if combs are extracted during the honey flow and then replaced in the hive they are virtually in constant use instead of being used as storage only. One of the pleasures of bee-keeping for honey production is to acquire the different varieties of honey. In many districts the early spring blossoms, particularly in woodland districts where wild cherry is abundant, sealed surplus of delightful honey can be obtained but only in limited quantities.

A bee-keeper having say three hives may average from this source perhaps two sealed combs of two and a half pounds each net weight, thus producing fifteen pounds of spring honey. Later, following the apple and hawthorn, the sycamore and gooseberry, the clover, lime and heather. In districts where a surplus is obtained from mustard charlock and the brassicas it is most important that extraction is carried out piecemeal as honey from these sources granulates very quickly and it is far better that granulation should take place in the jar than in the comb.

The man who would store all until the end of the lime and clover and then extract altogether, has an admixture which may be pleasant but which lacks the characteristic of the individual flavours.

One important advantage of extracting piecemeal is that when the combs are put back into the hive they are already fully drawn out and in addition are wet with honey which makes them readily acceptable by the bees for refilling.

In wooded districts there is the danger of honeydew, which if extracted with the bulk may result in a spoilt season's work. The chief danger in leaving all on the hive is the danger of losing the surplus either by robbing or inclement weather. Robbing, described elsewhere, is often the cause of the loss of not only the surplus stores but also of the bees themselves. Inclement weather at the close of the honey season has produced many a heart-broken bee-keeper. A fortnight of wet weather and as much as 30 to 40 pounds has been known to be consumed by the bees, but if the bee-keeper has taken his surplus he can tide over the wet period by feeding his bees with sugar syrup.

A disadvantage of extracting piecemeal is that extracting is often a messy business. Also when extracting takes place a certain amount of waste takes place by honey adhering to the sides and wire frames of the extractor.

The same amount of waste takes place for 50 frames as for 5, so that the more times during the season the extractor is used the more the total waste.

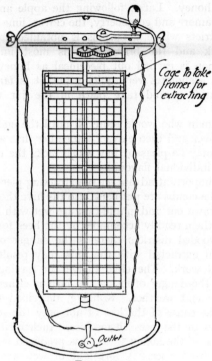

Cage to take frames for extracting

Outlet

FIG. 24
Simple Honey Extractor

Extracting Honey

There are many admirable mechanical extractors on the market and they vary in price according to their size and capacity. No small bee-keeper need go to the trouble of purchasing one if he is a member of his local bee-keepers association. Most bee-keepers associations own an extractor and a ripener. The principle upon which most extractors work is that the combs have the surface cappings cut off and are placed in the extractor, the honey being extracted by centrifugal force. Other extractors operate by having the frames placed in the cage of the extractor at right angles to the outer drum, the drum rotating and the honey being extracted by suction. The honey is thrown on the sides of the drum. It then runs down and is drained out of the bottom through a treacle tap.

Cleanliness is essential if the bee-keeper is to suceed in marketing his surplus honey for nothing is more likely to discourage customers from future purchases than honey having a dirty appearance.

Therefore, before commencing, scour out the extractor and ripener and all its parts with copius quantities of hot water. Wipe it out and set it before a warm fire to dry out thoroughly.

It is most important that before any honey is uncapped or exposed in any other way that the room in which the work is to be done should be made proof against bees and wasps ; otherwise large numbers will enter and impede the work. Never attempt to extract out of doors as the risk is not worth what the consequences will probably be.

It is unwise to expose any honey or syrup near any apiary otherwise robbing will probably ensue.

Next, take the crate of combs awaiting extraction and clear away from each frame any loose propolis. Hold the frame up to the light to see if there are any dark looking cells. Bees store honeydew in cells on its own and do not often mix it with honey. Remove any cells containing pollen with the point of the knife. Pollen does spoil honey.

A note on honeydew appears earlier in this book.

It may be that the dark cell contains pollen which the bees store for feeding the young bees. In no circumstances should this substance be allowed to become mixed with the honey as it will spoil its taste, making the honey taste strong and powdery. Any honey so contaminated should be fed back to the bees and they will distil it afresh. If the bee-keeper follows the advice given earlier and uses drone foundation in his extractor frames he will certainly avoid the pollen trouble as bees do not often store pollen in drone cells (see Plate II). Some readers have written the author saying they have found pollen in drone cells but the author has never seen it. There are two ways of dealing with honey-dew. First, if these dark cells are capped over the cappings should not be disturbed but the rest of the cappings over the normal honey should be removed and the frame is then ready for extraction. When extracted, the cappings over honeydew should be removed, the honeydew washed out with a syringe of warm water and the comb placed back in the hive. The second way is first to uncap and wash out the honeydew with a fountain pen filler, then to uncap the remainder and extract. The bees will then clean it up.

This is the simplest method of uncapping extractor combs. First, have ready a carving knife, a large jug of boiling hot water and a large meat dish or large pie dish. Stand the frame vertically in the dish, holding one lug in the left hand and the other lug standing in the dish. Then lean the frame away from you at an angle of 30 degrees. Take the carving knife, which should have been in the hot water for a few moments, in the right hand, wipe it and commence to cut the cappings by bringing the knife upwards in a sawing motion. It will be noticed that immediately under the capping wax there is a thin air space. In course of time you will become more expert and the knife will cut through this space very easily with the result that the whole capping will be cut away at one attempt with little or no honey adhering. Don't be easily discouraged and keep the knife blade hot, otherwise the comb will be injured by dragging. If you notice any dragging and collapsing of cells, stop the cutting and pick out the damaged broken away part and start again.

The uncapping should be repeated on the other side of the comb and when both sides are done the comb should be placed immediately in the extractor making sure that the wire net in the extractor is placed on the outside of the comb. This wire net is used in order to prevent the comb collapsing when the extractor is working.

It will be found that holding the frame at an angle of 30 degrees away from you will result in the wax capping falling away from the uncapped cells into the dish.

The extractor uses centrifugal force to extract the honey. The combs should all be fixed in the cage with the lugs in the corresponding position, in other words, when the combs are in the cage the lug ends or top of the frames should not be adjoining each other. This ensures that the cells of each comb are all pointing in a similar direction. The combs should be of approximately even weight to stop the extractor from wobbling when in action. The extractor should then be rotated slowly and it will soon be seen if it is being rotated in the correct direction. It should be rotated so that the cage is moving in the direction which is opposite to that in which the cells adjoining the wire net are facing.

After it is apparent that the operation is succeeding stop the extractor, take out and reverse the frames one by one so that the unextracted side faces outwards. Turn slowly for approximately the same number of turns as were given to the first side. It is important that the extractor should not be turned rapidly whilst either side of the comb is full of honey or the combs will break. The speed of the extractor may be increased in inverse proportion to the amount of honey left in the combs. The combs after being completely extracted should be put back in the hives. If, however, it is not intended to work for any more honey the frames should be placed in a crate and put back in the hive but over a Porter board which is illustrated earlier in this book. After they are placed on the hive the slide of the Porter board should be placed in the " open " position and left like that for about a week. In the meantime, the bees will have cleaned out the combs and taken down any remaining honey and stored it in the combs below the Porter board. The slide in the board can then be moved into the " shut " position.

In the next 24 hours the bees remaining in extracted combs will have joined the rest of the stock, making their way through the escape in the board. The combs can then be removed and stored for the winter.

" Ripening " the Honey

This is the generally accepted, although erroneous, term for storing honey prior to bottling. The ripener is a cylindrical container made from tinned sheet iron having a funnel filter at the top and a treacle tap at the bottom. When the honey is whisked out of the combs in the extractor it flies off the combs in small particles and often small pieces of wax go with it. The honey settles at the bottom of the extractor and is filled with minute bubbles and small fragments of wax. It is therefore run out of the extractor into the ripener and allowed to stand for several days before bottling. The filter at the top of the ripener is removable. It consists of a funnel with a gauze at its base. This gauze is only small enough to catch large particles of foreign matter so it is necessary for some other filtering medium to be put there in addition. A double thickness of muslin is adequate to catch smaller particles of wax, etc. This should be tied on the outside of the funnel. There is usually a flange round which it can be tied.

The honey is then run from the extractor into the ripener through the filter. This is rather a slow process, but it is well worth while. Nothing looks worse to a critical eye than cloudy honey with particles of dirt either suspended in it or lying at the bottom of the jar, and such honey never wins prizes even when granulated for dirt shows up even then. In spite of every care taken in cleanliness it is surprising to see the quantity of foreign matter which the filter will collect.

The honey should now be allowed to stand in the ripener for several days in a reasonably warm place. This enables the minute bubbles in the extracted honey to rise to the surface. Do not be in too great a hurry to bottle the honey otherwise disappointment will follow in that a scum of bubbles will appear in the top of the jar and so spoil the general appearance of the honey. So people run their honey straight into jars from the ripener. It is a matter of choice.

Wax Capping

The wax cappings may be dealt with in several ways. In the days when mead was more frequently drunk as a beverage the cappings and the adhering honey were steeped in water which was then allowed to ferment. Some bee-keepers place the pie dish which they have used to collect the cappings into a slow oven. This melts the wax which floats in a liquid form on the surface of the honey. The whole is then allowed to cool, the wax sets hard, is then cracked and the honey poured through the filter. This has disadvantages in that the heating of honey reduces its aroma. It may, too, become overheated in the oven and be ruined. Further, as the wax melts, a certain amount of pollen which is mixed with wax cappings comes into contact with the honey and tends to spoil it. The most satisfactory method of daeling with it is to place the whole of it in the top of the filter after having passed all the other honey through first. In two or three days time, the cappings will be almost drained clear of honey. They can then be washed in a little clear water which should be fed back to the bees. In this way not any honey is lost or spoilt.

Use may be made of a Solar Wax Extractor (see *Fig. 25*).

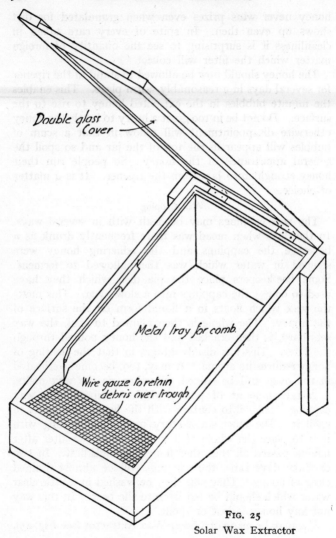

Double glass Cover

Metal tray for comb.

Wire gauze to retain
debris over trough

FIG. 25
Solar Wax Extractor

The sun's rays are used to melt the wax capping with this simple device.

Extracting Unsealed Honey

Bees do not seal over their honey until it is mature. By that it is meant that the honey will not be sealed until the surplus moisture has been evaporated from the honey.

There are two main objections to the use of unsealed honey and particularly to its sale. First, it is low in sugar content compared with sealed honey as it has too great a proportion of water in it and if sold the seller is liable to be charged with selling " to the prejudice of the Purchaser." Secondly, it is highly probable that it will ferment. Therefore, if you have any combs which are partly unsealed, extract the unsealed honey first and feed it back to the bees, for preference to a hive which is finishing off sections. Extracting unsealed honey is very simple. It means that you do not uncap the sealed honey but merely put the combs into the extractor as they are and the sealed honey will remain in the comb for extraction later.

Bottling Honey

After the extracted honey has remained in the ripener for a few days, it will be noticed that there is a scum of very fine bubbles on the top. This is perfectly pure and can be skimmed off with a warm spoon and used in the household. The honey is then ready for bottling. For household use, ordinary jam jars are quite adequate but for the sale of the surplus attractive jars help considerably in securing a market. These jars are of two types. The " tall " and " squat " screw top pound and half pound jars. The squat jars have the tendency to make the honey look a little darker and for show purpose are not very popular in consequence. Special jars with burnished aluminium caps are available for those who intend to compete at the show bench. Squat jars are much more popular with consumers as they may be used more readily on the table and are becoming more popular at shows.

Great care should be taken to see that jars contain the correct amount of honey. If you sell a jar as a 1 lb. jar, that jar should contain 1 lb. Tricks of the trade unfortunately have already stepped in and jars which hold only 14 ozs. have been introduced and take cover under the term " reputed

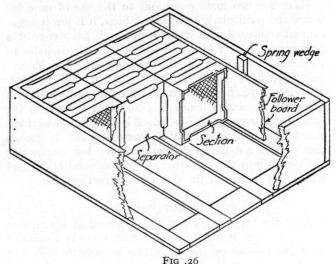

FIG .26

Details of a Section Crate

1 lb. jars." Weigh one or two jars empty and weigh them full and if all your jars are standard you will soon know how far up to fill them. It is better to err on the generous side rather than spoil your future clientele.

Don't sell all your surplus to one customer, the more people who buy your honey the greater will be the demand in the next season. Ask your regular customers to return their jars when placing their next order and a considerable sum will be saved in this way.

Section Honey

Section honey is a popular but hazardous method of producing honey.

There are many stocks of bees which will not for some reason or other take kindly to sections. In many cases it is found that rather than work on sections the bees will avoid completely the additional storage space which the bee-keeper places at their disposal, and will swarm particularly so if the first super placed over the brood-chamber consists of a crate of sections.

The usual section crate for a standard hive consists of 21 sections. There are three rows of seven in a row. The wooden section boxes are purchased in the flat in cases of 100. Great care must be taken in the assembling of the sections. It is desirable to make a "matrix" box into which the section can be folded. This box is made very easily. It is made just large enough to hold a section but has the top and one side missing. Take one flat section box, moisten the wood at the back of the grooves and fold over two of the sides at the grooves and slide into the matrix box leaving the split top with the dovetailed end upright. Take a sheet of thin wax foundation which is also purchased in packages of 100 and slide it into the grooves on the inside of the section box, making sure that the foundation is vertical and correct as shown in *Fig. 9*. Then fold over the split top and secure the dovetailed ends. After having done one crate in this way, set the sections with the split tops upwards in three rows of seven, each three being separated by a metal separator. These metal separators are used to prevent the sections being joined together with bracecomb. Bracecomb is honeycomb built by bees to tie together comb which is more than a bee space ($\frac{5}{8}$ths of an inch) apart. Bracecomb is often built by the English black bee which incidentally is usually a good section worker and makes a very white capping.

The best time to place a crate of sections on the hive is when the honey flow is well under way and a crate of shallow frames is almost filled with sealed honey. The bees will then have to pass through the sections in order to reach the shallow frames and in this way become accustomed to them. An inspection about seven days afterwards should be made to see if the foundation is being drawn out by the bees and at the same time an inspection should be made to discover any queen cells. If none of the section foundation is drawn out and Queen cells are discovered, one can be pretty certain that rather than use the sections they intend to swarm. In this case the section rack should be removed and the Queen cells also and a crate of shallow frames inserted in place of the section rack. Care should be taken to see that this stock does not continue with its swarming impulse by using the methods described in an earlier chapter.

A stock, however, will probably be found which will accept a rack of sections. If the honey flow is on then it is astonishing with what rapidity the bees fill the section boxes. It is no uncommon thing for a stock to complete a rack of 21 sections fully filled and capped over in one week. More often the process is slower. When several sections are seen to be capped over, remove the rack, take out the completed sections, move those already started to the middle of the rack and fill up the blank spaces with fresh section boxes and return the rack to the hive.

It is desirable when working for section honey to take away the sections as soon as they are sealed over because if they are left in the hive for any appreciable period after sealing they become " travel stained," that is, they become darkened in colour with the constant walking over by the bees. This makes them less attractive for the market.

Towards the end of the normal honey flow it is desirable to collect from the hives with sections all those which are nearly complete and place them in one rack on the best hive. If necessary, extract the partly filled sections and feed back the honey together with the washing of the cappings of the shallow bars, care being taken to do this at night-time and by means of a rapid feeder. In this way many sections will be completed which would otherwise be quite useless to market. Those which are extracted can be packed away with care for preference in a biscuit tin and used either in the following year or for the heather honey flow. If used in the following year they will be readily accepted by the bees who will start with partly drawn out combs which is such an advantage.

If the combs are used for heather honey they will in all probability be completed. The advantage of drawn out comb in honey production cannot be too strongly stressed.

It is a well-known fact that it is useless to hope to make a real success of extracted heather honey if the combs are to be retained for further use. Even when drone comb is used and the comb is put into the extractor warm it is a most arduous task, unless the extractor is power driven. So, for the average bee-keeper who seeks heather honey, the section is the better proposition. Heather for the purposes of this paragraph means, of course, Ling Heather (Calluna Vulgaris) and not Bell Heather (Erica Cinerea) the honey from the latter being readily extracted.

Sending stocks of bees to the heather is not a very difficult matter, many associations organise collective transport and arrange for suitable " out apiaries." The secretary of your local association will give you details of the arrangements which are available for its members.

The advantage of sending bees to the heather is that the honey season is prolonged, the bee-keeper receives an added return on his outlay and heather honey always commands a higher price than other varieties.

There are several well known hints regarding heather honey production. They are briefly as follows. Don't place the hive in a valley on the moors where mists hang

D

about or near heather growing on damp boggy land. Don't place your hives too near to those of others. You may pass on disease from your bees to the hives of others or you may find your bees infected by theirs. In any case too many bees in one locality tends to reduce the total amount of honey obtained from each hive, although saturation point is not reached unless there are very large numbers of hives.

The wax on heather honey is usually much whiter than on other honey and this makes it the more attractive as a marketable honey.

It is reputed that bees do not winter well on heather honey, therefore, when the bees are brought back they should be fed well with sugar syrup. For details of the method of feeding see the chapter on Winter Feeding.

In the South bees winter quite well on heather honey.

Skep Honey

At one time this was the principal source of honey production in this country. The method of obtaining the honey was both cruel and uneconomic.

The skeps which were heaviest were held over burning sulphur so that the fumes killed the bees which dropped, dead, out of the hive.

The combs which were left were cut out and outer combs were used as comb honey and the brood combs containing honey were bruised and broken, put in a muslin bag and hung over a container in front of a warm fire. This method resulted in the destruction of the entire stock of bees, the honey comb and the brood-comb and, further, the honey which was obtained was anything but pure, it being in part obtained from that stored in brood-comb where the faeces of the larvae had accumulated.

Later the process of driving bees out of skeps was adopted, with the result that the bees at any rate survived, although the combs were ruined.

The method of driving bees is as follows. Have available a spare skep with or without comb. Give a puff or two of smoke into the skep with bees. Too much smoke is undesirable. Turn the skep of bees upside down and place the spare skep over the exposed bees. Raise the front of the empty skep to an angle of 30 degrees and fix it in that position with driving irons. These irons are about 1 foot long and have spikes at either end, both at right-angles to the main iron and two of these are stuck into the skeps to keep them in position. The operator then beats the lower skep repeatedly with the palms of both hands for about five minutes. The bees object very much to this disturbance and quit their home and run up into the empty skep. It may be necessary to continue this procedure for longer than five minutes until all the bees have gone up. The new skep should then be placed on the stand of the old skep and fed with sugar syrup until there are sufficient stores for the winter. Fortunately, the present method of procuring extracted honey is simpler in some ways and much more hygienic.

Casual Honey

Occasionally a bee-keeper is asked to remove a stock of bees from the eaves of a house or some such place, where they are unwanted.

Sometimes phenomenal quantities of honey are obtained from these sources, dependent upon the number of years that the bees have been in occupation. The writer himself has tackled one of such stocks and six hours work produced 16 pailfulls of sealed honeycomb and one nucleus stock cut out of the brood-combs. The mass of combs measured 4 feet 6 inches across, were 2 feet 6 inches tall, tapering off to 6 inches. The main stock was left behind, together with about one-third of the total quantity of honey.

Certain experience was obtained which is worth recording. A great deal of honey was cut away in the comb without disturbing the bees to any great extent, but once disturbed they became extremely vicious, and even the best protection failed to stop dozens of stings. The bees, in large numbers, made for the hurricane lamp used.

The smoker was used extensively for several minutes on end and then the work proceeded with less interference. It was found helpful to have a few quarter of an hour breaks. This seemed to have the effect of settling the bees down again.

The honey collected was mostly granulated, having been stored in the comb for a considerable length of time. The comb was spotlessly clean and was all drone cell except those cells containing pollen. This confirmed the view that all storage combs should be fitted with drone foundation, as bees make drone comb for storage in their natural state.

The honeycomb collected in this way can either be packed away in tins or reduced to " run honey." As it was granulated in this case, it was put into a clean pail and put into a copper half filled with boiling water. After an hour the honey had melted out of the comb and the wax had also melted. The pail was taken out of the copper and set on one side to cool. The wax, solidified, was lifted off in a cake and the honey run through the filter into the ripener. Later, this was run off into jars and has proved a very good honey indeed.

Some of the meltings down were spoiled by some comb containing pollen contaminating the rest, but this was useful in supplying winter storage for another stock. The comb should have been held up to the light and the pollen would have been identified and could have been eliminated.

The nucleus was made by cutting out of the brood-comb pieces the size of an ordinary standard brood-frame and were inserted the correct way up in empty brood-frames and kept in position by each being tied round with red tape. Care was taken to ensure that there were freshly-laid eggs in one comb and all went well. The old stock was left in the roof with substantial stores and will be available for honey gathering later on.

CHAPTER VIII

INCREASE OF STOCKS

THE bee-keeper must know how to increase the number of his stocks, for there are seasons when disease overcomes some and in some years stocks die out through lack of winter stores.

In due time and from experience the bee-keeper will learn which is his best all round stock and from this best stock he should provide the Queens which are to populate his other hives.

There are five main ways of increasing stocks. First, collecting swarms; secondly, Artificial Swarms; thirdly, Formation of Nucleus Stocks; fourthly, Nucleus Swarms; and lastly by the utilisation of a number of stocks.

Collecting Swarms

This method is most precarious. Modern bee-keepers deprecate the practice of swarming. The accumulation of large stocks of surplus honey cannot be expected from hives which swarm however early in the season that may occur. Also, unless the bee-keeper is available at all times to deal with the swarm, his efforts in encouraging the stock to swarm may be lost as the swarm may abscond in his absence. If, however, the bee-keeper intends to rely on casual swarms his chance is even more precarious. He may collect a number which may or may not be useful. One thing is certain and that is that he will have no idea of the characteristics of his bees until the end of the season. He may be blest with good fortune and yet he may acquire diseased stocks which may infect his others and so be a danger to his apiary and those of his neighbours.

There are in many parts of the country "bee-men" who are called in to remove swarms. They hive them in skeps, boxes and derelict hives. They collect a certain amount of skep honey and occasionally section honey by placing a crate of sections on the box or skep. They care nothing about the welfare of the bees and are unaware of their possible diseased condition. Such men are a source of danger to the community of bee-keepers. Don't in any circumstances buy bees from such bee-keepers.

The principle objections to collecting swarms are that the bee-keeper collects bees whose swarming propensity has been proved and that a swarm usually consists of old bees with an old Queen.

Artificial Swarms

This is by far the simplest, convenient and most effective way of increasing stocks but should be carried out early in the season if surplus honey is to be expected. One great advantage is that the Queen who will operate in the artificial swarm for the rest of the season is the Queen whose characteristics are already known.

The following preparations are necessary to make a real success of the operation. First stimulate the stock which you intend to operate upon. This stock is called the parent stock. There are two methods of stimulation as described earlier under the heading of " Spring Stimulation." Briefly this consists of stimulation which it will be remembered is carried out in two ways:— stimulative feeding and brood spreading. The feeding should be commenced about the first and second weeks in April and the capped honey on the centre brood-combs should be uncapped or bruised so that the bees will use these stores and fill the empty cells with brood. A regular quantity of thin warm syrup should also be given. If these instructions are carried out with care it will be found that the stock will be extremely strong by the first week in May. It will be full of bees and brood and it may be found that there are signs of Queen cells.

Preparing a strong stock is the first important step to the sucess of the whole operation.

Have ready an empty hive with the brood chamber ready to receive the artificial swarm. It should be empty and ten spare brood-frames should be available to fill up the blanks which will be created in the parent stock and also to act as brood-frames for the artificial swarm.

Lift the parent stock about four feet to one side and place the new hive in the exact position of the parent hive. Open up the parent hive and find the Queen. Take the brood-frame on which the Queen is and place it in the new hive. Take a comb of sealed brood, too, with the adhering bees and place it next to the Queen. Then take two frames containing honey and pollen but no brood or eggs. These are usually found on the outer combs of the parent hive. Place one on each side of the two frames already in the new hive. Fill up the remaining space in the new hive with the spare brood-combs three on each side of those already put in the new hive. This will make the full complement of 10 brood-combs in the new hive. Place a quilt over the 10 brood-combs and pack the hive down and leave it alone for several days. Now as to the parent hive. The frames containing brood should now be placed together. It will be remembered that four frames were removed so that it will be necessary to place the last four spare frames in the parent hive. These should be put on the outsides, two on each side. The hive should then be closed down. The parent stock should be fed for a few days as much of its reserve stores have been removed to the new hive. Also, the bees which remain in the parent hive are young bees and have not yet become foragers.

The result of this operation is that the flying bees will return to their original situation and join their own original Queen. They will soon assist the Queen to build up the new stock. The young bees will soon hatch out of the sealed brood and will act as nurse bees, leaving their empty cells ready to receive the Queen's eggs. The bees in this time will build up to produce a strong stock and will all being well produce surplus honey the same season. The parent stock in the meantime will have realised the loss of their Queen and will proceed forthwith to make a number of Queen cells using the eggs which the parent Queen had laid in worker cells. In due course a virgin Queen will be hatched out and after a lapse of about four weeks from the operation the stock will begin to build up again and will be strong enough to produce surplus honey. One big advantage will be apparent, that is, the parent stock having been reduced at a time when swarming was imminent will have lost its swarming instinct for the season. Another advantage is that the parent stock will go forward to the following season with a new Queen.

Formation of Nucleus Stocks

Again as in the case of artificial swarms only strong stocks should be used and the method of stimulating described earlier should again be carried out. The operator should in the case of forming a nucleus wait until a little later in the season than in the case of making an artificial swarm, the reason for this being that in the case of an artificial swarm a strong swarm with existing brood, flying bees and a laying Queen is able to carry itself on just as well as if it were a natural swarm, whereas in the case of a nucleus there is no Queen to carry on the laying and a great deal less number of flying bees to forage for and build up the stores.

The operation is carried out as follows. Prepare a spare hive ready to receive the nucleus or alternatively a nucleus hive. The illustration of a nucleus hive shown here is a very useful part of a bee-keeper's equipment and every bee-keeper should make or acquire one capable of holding four or five brood-frames.

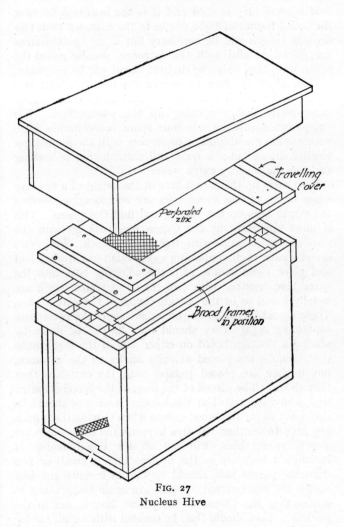

FIG. 27
Nucleus Hive

If a spare hive is used and it is the intention to have the brood-frames at right angles to the entrance then two division boards will be necessary but if the brood-frames are placed parallel with the entrance, usually called the warm way, then only one division board will be necessary.

First place the hive to receive the nucleus about two yards away from the parent hive and have all the equipment ready before opening up the parent hive. The equipment should include four spare brood-frames filled with worker foundation or preferably with the foundation already drawn out; a quilt and extra felting or sacking to keep the nucleus extra warm.

Next open up the parent hive at the height of a fine day when many of the flying bees are out foraging, leaving the young bees on the combs, and find the Queen. This is most important as the Queen must not go with the nucleus but remain with the parent stock. Find two brood-combs each containing eggs which are freshly laid and place these combs with the adhering bees into the spare hive together with two combs of stores which are usually found to be the outside combs of the parent hive. These should be placed one on either side of the two combs containing eggs. They should then be packed down by placing a division board on either side of the four combs if the combs are placed at right angles to the entrance, but if they are placed parallel with the entrance then they should all be placed at the front of the brood-chamber with a division board at the back. Great care should be taken not to jar the brood-combs when moving them from one hive to another. If this happens the bees, which are mostly young bees, will fall off the brood-combs. If insufficient bees are in the nucleus then a comb or two from the parent hive should be shaken in until the four combs are well covered, special care again being taken to ensure that the Queen of the parent hive is not in the nucleus. They should then be covered with a quilt having a feeder hole cut in the top; a rapid feeder should be placed over the whole and a supply of warm syrup should be given daily to the nucleus to ensure that it will not perish of hunger. It should again be emphasised that whatever is given to the bees at this stage by way of syrup will be amply repaid by a stronger stock later on. The entrance to the hive should be closed down to two inches and should be very lightly plugged with dry grass. This enables the bees to breathe through the grass but prevents their escape back to the parent hive. This should be removed after 48 hours during which time the bees will have lost their homing instinct and will regard the nucleus as their home.

When the bees discover that they are Queenless they will immediately commence to form Queen cells to create their new Queen. It will help matters very much if a Queen cell either from the parent stock or from another good stock can be inserted in the nucleus. Usually it will be accepted by the bees and they will continue to rear it to fruition. This means that valuable time will be saved if a sealed Queen cell is inserted. Recently, in early June, the writer opened a hive for a friend some seven miles from his apiary and discovered a number of excellent Queen cells sealed. As they were being cut away in any case to prevent swarming he placed one in a match-box and then in his waistcoat pocket to keep it warm and returned to his apiary, made a nucleus and inserted the ripe Queen cell in the nucleus by wiring it with tinned wire to one of the nucleus combs. Next day, on inspection, it was noted that the bees had annealed the cell to the comb with wax and had freshened up the appearance of the cell. The following day, on inspection, it was noted that the Queen had hatched out and was accepted by the bees who formed no other Queen cells. In a further ten days, eggs and unsealed brood were discovered. This nucleus did well and produced a small amount of surplus honey.

An excellent way of starting a nucleus going is to purchase a young fertile Queen from an apiary of repute. This arrives in a Queen travelling box [*Fig. 28*].

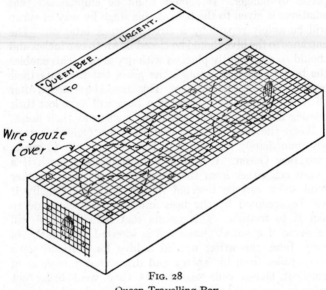

FIG. 28

Queen Travelling Box

Normally a nucleus if made before the end of May will build up to full strength before the honey season is over and may even produce surplus honey but the help that is given at the start both by careful feeding and the provision of a ripe Queen cell will be reflected in the final result.

Nucleus Swarms

There are two ways of carrying out this operation both of which in practice meet with equal success.

First method. Carry out the formation of a nucleus as described earlier but instead of packing the four brood-frames together and shielding them with division boards, place six brood-frames with either foundation or drawn out comb, for preference the latter and instead of putting the nucleus on a new stand put the parent hive at a distance and the nucleus on the stand of the parent hive.

In this way the nucleus will receive a large number of flying bees and build up very rapidly but the parent stock will be depleted rather heavily although by way of compensation will have a laying Queen and a fair number of young bees to help build up. If this process is adopted the parent stock should be fed also as the foraging bees to a great extent will have joined the nucleus.

Second method. In this case, two strong stocks are needed. A nucleus should be formed of two combs of eggs, and sealed brood and two combs of food from the parent hive. Six brood-frames with foundation or drawn out comb should be added. The first parent hive should remain on

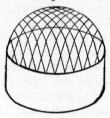

FIG. 29
Wire Gauze Cage

its own stand but the second parent hive which is to supply the flying bees should be removed from its stand and moved at least 3 yards away and in its place should be put the nucleus. Thus it will be noted that the first parent hive supplies the eggs and sealed brood, but the second parent hive supplies the bees.

This is an extremely good method for both parent hives rapidly overcome their losses and the nucleus will forge ahead with the help it has received. Only the nucleus need be fed and then to only assist the bees to draw out the comb foundation.

A nucleus can always be strengthened by transposing the nucleus and the stock from which flying bees are to be taken, but when a nucleus has hatched out its Queen, a special precaution is necessary. Open the nucleus, find the Queen and place a wire gauze cage [*Fig. 29*] over her with two or three young bees with her. Add additional brood frames and transpose the nucleus and the other hive. The nucleus will receive the flying bees from the other hive which, in its turn, will receive the flying bees from the nucleus. After 48 hours the Queen will be accepted by the bees and should be freed from the cage. The young bees are encaged with Queen to ensure that there are bees to feed the Queen who may not be capable of feeding herself. A young Queen not in lay can certainly feed herself and there seems no reason why a Queen in lay could not do so, although admittedly she does not do so, being tended by the workers instead.

The great advantage of forming nuclei from strong stocks is that the bee-keeper is assured that because of their strain the nuclei will also prove themselves by becoming strong stocks, also that the withdrawal of part of the strength of strong stocks early in the season tends to overcome the swarming impulse before the main honey flow arrives.

Using a number of stocks

This method results in less drain on the parent and other hives than any other methods. Take, for example, five hives. Take two combs of eggs without bees from one, two combs of sealed brood without bees from the second, two combs of sealed brood without bees from the third, the same from the fourth, pack these with two combs with foundation one on either side, making ten combs in all.

Place the whole on the stand of the fifth stock, which is moved to one side. The stocks which are depleted have the blanks filled up with fresh combs or frames with foundation. In this way one hive supplies the eggs, three supply the sealed brood which will, in turn, supply the foraging and nurse bees for the new Queen and the fifth, which supplies the bees.

Within a very short time this will be a very strong stock indeed. There will be ample room for the new Queen to lay as all the sealed brood will have hatched out by the time the new Queen is laying. This stock is almost bound to produce surplus honey. The other stocks, too, will recover quickly and will scarcely appear to be affected by the operation. Almost certainly the swarming impulse will have been controlled for the season, particularly if the operation is repeated three weeks later, and another nucleus formed in a similar way.

This is the method usually adopted by bee-keepers who make stocks up for sale. It pays the beginner to purchase a stock of this type because he is assured of a young Queen in a stock which will prove a honey gatherer.

There is one other method which permits of an increase of stock and yet ensures a surplus honey yield. If in spite of all precautions taken by the bee-keeper a swarm emerges, he should follow the following instructions. It must, of course, be presupposed that one of the methods of swarm control which he has adopted is the supering of the parent stock. The swarm should be hived temporarily in a skep or box. A new hive should be prepared and placed immediately adjoining the parent stock. In the evening the swarm should be hived in the new hive and the super of the parent stock placed over the brood chamber. The parent stock should be closed down and fed as it will have lost the greater portion of its stores, and on the morrow will lose many more of its bees for reasons explained. At the height of the following day the parent stock should be removed some distance away and the new stock moved slightly so that it stands in a position which is just midway between where the two hives were placed the day before. In this way the swarm has added to it additional flying bees. It will be some time before all the foundation is drawn out on the ten frames in which the Queen will ultimately lay. During that time as all the bees are flying bees, a substantial amount of honey will be stored in the super. If this is done early in the season, the swarm will prove a good surplus honey producer.

The drawback to this method is that as the parent stock is very depleted owing to the departure of both the swarm and, later, the flying bees, it is a longer time recovering and cannot be expected to be a surplus honey producer. However it is a maxim among bee-keepers that one strong stock is worth two weak ones, and this practice is strongly advocated.

CHAPTER IX

THE FEEDING OF BEES

THIS can be classified into three principal groups. Spring or Stimulative Feeding, Summer Feeding and Autumn or Storage Feeding.

The purpose of feeding may, to the beginner, seem strange as bees are reputed to be producers of sweetness rather than consumers of it.

Spring Feeding

Spring feeding should not normally be necessary but in the present period of shortage of sugar it is not possible to feed adequately in the autumn to render spring feeding unnecessary.

Feeding at any time always stimulates foraging.

The purpose of Spring feeding is to encourage the bees to breed in greater numbers and at an earlier time than they would normally breed. Earlier in this book is a description of Spring feeding. Before supplying any syrup look down between the combs. Do not disturb the bees unless it is a warm day. See if there is any quantity of sealed stores particularly at the centre of the cluster. If this is visible uncap the sealed honey there either with a penknife or in some other way so that the hive is little disturbed. It is generally accepted that if bees are disturbed too greatly at this season that they frequently " ball " their queen. That is, they cluster round her so tightly that she is suffocated. The bees will clear up and use the honey in the cells which are uncapped and make ready the cells for the Queen for immediate use. Further use may be made of this method until it is seen that four or five combs are being used for breeding. Then comes the time for the stimulative feeding by the slow feeder. Although the recipe given below is generally accepted as correct, the bee-keeper is advised to try spring stimulation with the very thin syrup which the writer has by experiment shown to be very effective. In any case recipes vary to such an extent that the life or death of the bees does not appear to be materially affected.

Digges advises equal weights of water and sugar ; that is equivalent to $1\frac{1}{4}$ lbs. of sugar to each pint of water. The author of the Ministry of Agriculture's Bulletin No. 9 on Bee-keeping advises half a pint of water to every pound of sugar, that is 10 ozs. of water to 16 ozs. of sugar.

The amount which is given to a stock depends upon its strength and also upon the amount of the stores existing in the hive but the syrup of equal weights of water and sugar can be given at the rate of a cupful every other evening. An occasional inspection will show whether or not the bees are storing the syrup in the comb. If they are storing then cut down the supply of syrup and resort to brood spreading.

Spring feeding is the feeding which appears to show the best results. The stocks grow more rapidly and the stocks get into the condition for which the bee-keeper has been longing during the winter months.

Care should be taken to feed the bees only in the evening and to keep the entrance of the hive closed down to three inches. The sugar used must be clean white sugar. Brown sugar gives the bees dysentery. The sugar must not be burned when dissolving in the water. There are those who aver that beet sugar is unsuitable for feeding bees and that only cane sugar should be used. This is, of course, absurd, as both the sugar beet and the sugar cane produce identically the same chemical substance which chemists call " cane sugar." A reason for this belief may have arisen, for in the early stages of beet sugar production Barium salts were used in the refining process. Barium is a cumulative poison and although great care was taken some very small amounts may have escaped with the product. These amounts were not sufficient to do any serious harm to human life but it is possible that even a minute quantity of a Barium salt may have been harmful to bees.

Barium salts are not now used in the refining of beet sugar which may be the reason for the present beet sugar being quite harmless.

There are those too who say that the syrup must be boiled to sterilise it. This too is totally unnecessary. Is the nectar given up by the flowers boiled? Why should sugar fit for human consumption have to be boiled for bees? The only virtue in boiling the water is that the sugar dissolves the more readily in it. Some add vinegar, some salt and others even beer and wine. What a waste of good material!

Summer Feeding

Many bee-keepers fail to make use of summer feeding. There is very often after the early fruit blossoms an hiatus before the flowers commence their nectar flow. This means that any surplus which the bees may have stored from the early blossoms may be completely used up in maintaining the growing stock. Not only may they have used that up but may even throw out of their hive the immature nymphs and grubs. This occurs particularly if the weather is bad and the bees cannot forage. Then is the time that the judicious bee-keeper will feed his stock liberally so as to maintain his stock in a strong condition and even increase its strength for the honey flow which is to follow. Care, however, should be taken to ensure that the bees do not commence to store the sugar syrup in the supers. Some take the supers off when they find it necessary to begin summer feeding to avoid storing syrup in the combs, but this is unwise for almost without warning the honey flow starts and the bee-keeper may lose a few days of valuable honey gathering He may, too, by his action begin to crowd out his bees with the result that the swarming instinct may be roused with unhappy consequences.

To avoid any trouble whatever with the storage of sugar syrup in the supers the perfect bee-keeper should feed with pure honey diluted with one cupful of water to each pound of honey. They can store this if they wish in the supers without harm.

To those bee-keepers who intend to send their bees to the heather a word of warning is due. When taking off the honey at the end of the clover flow, that is during the third or fourth week in July, do not rob the bees of all their honey. The end of July and the early part of August is often a very treacherous time and the stock may be totally unfit for the heather unless sufficient stores are left or unless generous feeding is adopted. Always be generous with your bees; they are generous to the bee-keeper.

Again, when taking all the surplus honey off the hive at the end of the honey flow give the bees a little syrup and keep it up until the main autumn feed in September. If this is done, the bee-keeper is assured of a very strong stock for wintering.

The syrup recommended for summer feeding, 2 lbs. of sugar to 3 pints of water which is approximately the strength of nectar which is drawn from the flowers. Nectar varies in its water content from approximately between 50 per cent. to 80 per cent. water, but may vary even more than this.

Autumn Feeding

It is often incorrectly stated that the purposes of Autumn feeding are twofold, (a) to stimulate the bees to breed and so go through the winter a strong stock, and (b) to have sufficient stores to live through the winter.

The second only of these purposes is correct. In support of (a) above, many experienced bee-keepers advise spreading the period of autumn feeding over a period of two to three weeks. This is wrong and wasteful. If the bee-keeper will only continue with an occasional small feed after the honey flow has ceased the stock will remain strong until the middle of September. Then the autumn feed should be given as rapidly as possible and it will be found that the bees will seal it quickly, breeding will be restricted as the available cells will be filled with stores. At Rothamsted Experimental Station better results are obtained by completing Autumn feeding by the middle of September at the latest. It should however be noted that Rothamsted is not a heather district. The restriction of breeding at this stage is not undesirable if the stock is strong. The writer recently made a feeder capable of holding ten pints of syrup. This was emptied by the bees in two days, filled again, and emptied this time in three days. An inspection later showed the combs very well filled with almost all the syrup sealed. The stock wintered well, was expanded to 16 brood-combs in the Spring, a six frame nucleus was taken, then a four frame nucleus 14 days after and three weeks later it threw a 4¾ lbs. swarm and gave 16 lbs. of surplus honey. This was all carried out in an experimental hive, details of which the writer proposes to record at some early date.

If the Autumn feed is given slowly the bees will believe a honey flow has started and breeding will be stimulated to an excessive amount and the stores intended for the winter will be consumed in feeding the grubs and young bees instead of being kept for winter feed. The correct method is to keep the stock strong all the time and then give the Autumn feed rapidly.

Use a rapid feeder, the larger the better. Feed if possible all your stocks at one time as this tends to avoid robbing.

The strength of syrup for winter feed should be greater

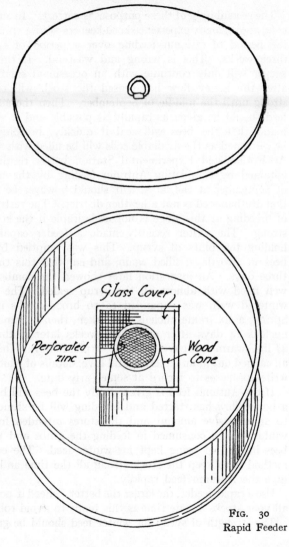

FIG. 30
Rapid Feeder

than for either Spring or Summer feeding. Two and a half pounds of sugar to one pint of water makes a very practical syrup. Some add oil of thyme and some chemicals to prevent fermentation but if a thick syrup is fed, and fed rapidly to a strong stock the bees will seal it over. Sealed syrup does not ferment ; unsealed syrup may, but the bees will always consume the unsealed first before breaking into their stores.

Candy for Winter Feeding

This substance is an unnatural food for bees and should be used only as an emergency measure. To buy it is expensive, to make it is found difficult by many and its use is a sign that the bee-keeper has doubts as to whether he has given his bees sufficient winter stores. Some bee-keepers of wide experience state that rather than give candy to their bees they would prefer to put on the rapid feeder and give the bees a dose of thick warm syrup, the reason given is that candy is worked at by the bees continuously and that disturbs the hive, whereas a feeder full of syrup is taken down overnight and the bees settle down again to their winter conditions.

A recipe which has proved satisfactory is given here. Three pounds of sugar, half pint of hot water and half a teaspoonful of cream of tartar. Place the water in a pan, heat it and add the sugar gradually, stirring all the time to ensure that all the sugar dissolves and does not burn. However slightly burnt, the mixture as it is will be useless for the bees. Add the half a teaspoonful of cream of tartar. The whole should be brought to the boil and allowed to boil for two minutes. Then stand the pan in cold water until the contents of the pan begin to cloud. As soon as this cloudiness appears stir hard and pour into the containers. The containers should be made so as to contain about 1 lb. of the candy. They can be made of a section box glazed on one side. This has the advantage that an inspection without disturbing the bees can be made. If the bees have upon inspection removed the greater portion of the candy, another box can be placed over, say, half of the feed hole by moving the original box half way off the hole. When the first one is all used up, it should be removed and the second one made to cover the hole, or a third one to replace the first. This may be continued and fed to the bees so long as they continue to remove it. The weakness of this method is that the bees remove the candy before consuming their own stores and even store the candy in their combs after liquefying it. The result may be that when the Spring arrives, the hive is crowded with surplus food ; it is a waste from the point that the effort and the candy have been unnecessary. The alternative to candy feeding is to ensure that the bees are adequately fed before being bedded down for the winter. A good method of ensuring that bees have sufficient food is to inspect the hive during a warm day in the early part of October. If, on average, the ten bars of the brood-chamber hold more than half their cells filled with sealed stores then there is sufficient to see the stock through the winter. One brood-comb filled with sealed stores contains about five pounds weight of honey or syrup. Therefore ten half filled brood-combs contain approximately twenty-five pounds of stores, which is regarded by experts as enough to see the bees through the winter.

If on inspection it is found that there is a deficient quantity of stores, put on a rapid feeder and give the bees sufficient thick syrup to make up their requirements.

It is important to remember that it is much better to err on the generous side than to lose a stock of bees through starvation.

It has been found after long trial at Rothamsted that the best results are obtained if each colony is left 25–30 lbs. of natural stores in the Autumn and then are fed with an additional 5–10 lbs. of concentrated syrup as quickly as possible so that it is stored and not used for immediate consumption. It is found that 35–40 lbs. of stores in all is quite sufficient to tide any colony over the non-productive months of the year without any further feeding whatsoever. Some experienced bee-keepers think even this is inadequate and the beginner is advised to be generous with his bees.

CHAPTER X

WINTERING OF BEES

Bee Passages

A VERY important matter in winter in the hive is the ability of the bees to reach their stores. Many stocks have been known to perish through their inability to reach their stores. Three effective methods of dealing with the problem are employed. The crudest method is to cut a hole about 1 inch diameter at the top of each comb just where it joins the cross bar of the frame. In this way the bees can get through to any comb in the hive for stores. This is not a satisfactory method for the bees will repair the damage with drone cells to the disadvantage of the hive. The second method of making a free passage is the placing of two sticks about a foot long and $\frac{3}{8}$ of an inch in diameter about two inches apart across the top of the brood-frames. The quilt is laid across over the sticks and this makes a kind of long tent through which the bees can pass to any comb for stores. The third and very effective method is to use a Porter board. Remove the bee trap and cover the hole with a piece of glass or zinc gauze. The Porter board by being set in a frame permits of the passage of bees to any part of the hive. Some claim this to be the ideal winter cover for bees and in support state that in natural conditions bees live in holes in trees which have a wooden roof and that a cloth roof is less natural. The advantage of having a piece of glass over the bee escape hole is that an inspection can be made without disturbing the bees. The piece of glass has one disadvantage in that condensation is likely to occur and drops of water may drop on the bees. This they hate so that the zinc gauze is advised for preference. If it is found necessary to feed the bees either with candy or syrup, this can be done by sliding the piece of glass or zinc gauze to one side. The writer has used this method quite successfully and without loss of stocks. If the zinc gauze becomes clogged with propolis it should be removed and scraped clean at the end of October, then replaced and will probably not be clogged up until the Spring.

Winter Packing

Coincidental with winter feeding is the matter of packing bees down for the winter. There are many and varied views on this matter. Some bee-keepers make special boxes containing chaff or cork granules and place this immediately over the brood-chamber. Others pack lots of felt sacking and old clothes over the brood-chamber. But all that is really necessary is the ticking quilt, a square of carpet underfelting and a folded sack. Cold of itself does not kill bees in the winter. But cold and damp will kill them. It is therefore more important to ensure that the hive is kept dry than to see that the bee-hive is overpacked with blankets of all types.

It has been recommended that the feed hole flap in the quilt be turned back and the hole covered with a square of perforated zinc and that this should be covered with a blanket of felt. The object in this is to see that there is sufficient ventilation in the hive to ensure that it is kept completely dry. The writer followed this recommendation but discovered that the zinc perforations soon became completely sealed up with propolis and that the bees had even dragged the hairs of the felting through the perforations and made the holes one mass of propolis mixed with hair. The following Spring it was found impossible to remove the felt without also moving the zinc.

Very little more winter covering is needed than summer covering. The writer finds that summer cover plus a folded dry sack is sufficient.

Winter work in the Apiary

Apart from Autumn feeding and packing down for the winter, there are a good many other matters which require attention. This is the time of preparation for the Spring and later the honey flow.

Storage of combs. All the shallow combs should be inspected by holding up to the window or bright light for traces of the devastation of the wax moth, they should be scraped clean of propolis and brace comb. The metal ends should be cleaned up and the runners of the shallow bar rack should be well vaselined. The crates should be sealed top and bottom with newspaper after placing a few of the volatile crystals of Paradichlorobenzine, commonly called P.D.B., in each crate. This acts as a fumigant against wax moth in stored combs. The crates should then be packed one on top of the other and stored in a place with an even temperature. Each crate should be labelled as to its contents stating whether the combs are worker or drone foundation. These crates should not be opened until they are required for the hive in the Spring where they may be wanted at a moment's notice. If they are labelled, mistakes such as placing drone combs over the brood-chamber as additional breeding room will be avoided. Queen excluders should be scraped and placed in between the wrapped shallow frame racks. This has two advantages. First, it keeps the excluders perfectly flat and secondly, it prevents any mice which may get into the stack of crates from passing from one crate to the other.

The roofs of the hives should be inspected to see that they are watertight. Any showing signs of leaks should be repaired immediately and damp felts replaced by dry ones.

The hives should be tilted forwards by raising the rear legs of the hive about one inch. This ensures that any rain driven in or condensation formed within the hive can readily run out.

In exposed districts the danger from snow is a serious menace to bees. A warm sunny spell after a fall of snow entices the bees out. The reflection of the sunlight on the snow gives the bees the idea that the Spring has arrived at last. They fly out in large numbers but the cold is too much for them, they land on the snow, become numbed and die. It is therefore desirable to screen the entrance of the hive when snow is lying on the ground.

The metal parts of feeders should be smeared with vaseline to prevent rusting and stored away until required.

The lifts from the hives should be removed, the paint-work brushed down with a wire brush and each lift given a coat of paint. Care should be taken to see that the top edges of the lift are not painted as this may become slightly sticky in summer and may result in jolting the hive when manipulating. This will disturb the bees unnecessarily and may anger them.

Repairs to hives and equipment should be carried out in readiness for the oncoming season.

Winter is a useful time to make spare equipment. Whatever you make see to it that it is standard in size. All equipment should be interchangeable. Nothing is more upsetting than to find out halfway through the manipulation of a hive that the equipment one is about to use does not fit.

If you make a spare hive, make it exactly like the others you have. This is both practical and economical.

Some precautionary treatment against Acarine disease (see chapter on bee diseases) might with advantage be carried out. The " tube wick " method described should be put into operation in the Autumn of the year.

CHAPTER XI

GENERAL ADVICE AND POINTS OF ETIQUETTE

ALL bee-keepers should be members of the local association. There are many advantages in so doing. Usually the local association provides a group insurance scheme against bee diseases and an insurance scheme against third party liability. Many associations provide arrangements for sending bees to the heather. Indoor meetings and lectures of value are arranged for the winter months whilst visits to and demonstrations in notable apiaries are arranged in the summer. Many associations arrange a " mart and exchange " among their members, also bulk purchase schemes. Lists are kept of those having bees for sale and those requiring bees. They also arrange a marketing scheme for surplus honey and collect beeswax for transmission to foundation manufacturers.

If you keep your bees in your own garden be neighbourly to your neighbours. Remember that a comb or jar of honey occasionally will sweeten them and you will not be unwelcome when your swarms lodge in their apple trees.

Don't keep too many stocks of bees in a suburban garden. Two or three is a reasonable number and no-one can take exception to it, but 10, 15 or 20 would be unreasonable and likely to lead the bee-keeper into trouble with his neighbours. In any case, it is inadvisable to keep more than two or three stocks in the first season and afterwards any additional hives should be kept in an out apiary where supplies of nectar are in greater abundance than in populated districts.

Don't be afraid to enter your products in the local agricultural and horticultural shows. This has proved one of the best marketing mediums and a means of learning a great amount about the craft.

...ve exposed an empty hive but fitted with brood-
...amber and old combs. This is "not done" amongst
...od bee-keepers because if a swarm from another
...ekeeper's hive should emerge and the scouts from the
...arm find this desirable home, then the swarm may
...camp before the rightful owner has time to hive it.

Manipulation of a Hive

When opening up a hive great care should be taken to
...sure that the bees are as little disturbed as is possible.
...aturally the apparent breaking up of their home does
...use the bees a certain amount of distress and concern
...d this is shown in two ways. Either the bees will
...ecome angry and will fly out and attack the operator
...r they will make for the honey cells and gorge themselves
...ith honey. The latter they do as a matter of instinct,
...r they will need that honey if they are compelled to
...nd a new home.

...Rough handling of the hive makes bees angry and they
...ill attack the operator.

The illustration, Plate III, shows the author's protection
...hen handling strange or "active" bees.

This method of approach has been found most satisfactory
...y the writer. First, have everything you intend to use
...n the hive handy and immediately available. Light
...he smoker and give the hive three good puffs of smoke
...irect into the entrance. Then, without hurrying, remove
...the roof and the outer lifts so that the interior is easily
...accessible. Do not jolt or jar the hive in any way.
...Remove the blankets, raise a corner of the quilt and give
...a small puff of smoke into the exposed combs. Pull off
...very gently the quilt and usually the bees will have their
...heads stuck into the honeycomb. Then take a "stink
...cloth," that is a cloth soaked in a weak solution of carbolic,
...lysol or Jeyes' fluid, and cover the exposed brood-combs
...Leave it there for a few seconds and then remove it. The

PLATE III.

Protection against stings: wire net veil, gauntlet gloves
and cycle clips on trousers.

The correct way to hold a comb, i.e., vertically over the hive. Note lift on left being used as a rest for comb.

The incorrect way of holding a comb. The comb should not be held horizontally and should be held over the hive.

PLATE IV.

If you have produced surplus honey
summer and sold it, you will probably be as
if you have any to sell. Keep a list of th
addresses of those people, you may need th
customers if your apiary develops.

Always be willing to send samples of you
agricultural centre when requested for tests
This will help you and help all other bee-keepe
to prevent the spread of bee diseases. The
ment, Rothamsted Experimental Station,
Herts., will help you if you are in any doul

If you develop a good strain of bees let your
friends know when you have a spare Queen cel
Never charge bee-keepers for Queen cells. If
the stocks of your neighbouring bee-keepe
probably in the end help you to maintain
quality, for who knows but that a drone fro
bouring bee-keeper may fertilise your next vi

Always be willing to look after the stocks
bee-keeper if he is ill or called away on importan
do not expect any reward for this, you yoursel
help some day.

If you are called in to remove a swarm fro
near another apiary, take the swarm, but before
it go to the other bee-keeper to ascertain if he
and if that be the case the rightful owner should
to hive the swarm. If, however, you are th
owner, you should proffer some good evidence
your swarm.

There is an ill-founded notion amongst bee-kee
it is permissible to trespass in pursuit of their
A case was, however, recently decided in the H
from which it is clear that such notion is now
effect. Keep sweet with your neighbours and
not meet with difficulty over trespass.

Some bee-keepers in the hope of catching stray

PLATE III.

Protection against stings: wire net veil, gauntlet gloves
and cycle clips on trousers.

The correct way to hold a comb, i.e., vertically over the hive. Note lift on left being used as a rest for comb.

The incorrect way of holding a comb. The comb should not be held horizontally and should be held over the hive.

PLATE IV.

If you have produced surplus honey the previous summer and sold it, you will probably be asked by others if you have any to sell. Keep a list of the names and addresses of those people, you may need them as future customers if your apiary develops.

Always be willing to send samples of your bees to the agricultural centre when requested for tests for disease. This will help you and help all other bee-keepers in tending to prevent the spread of bee diseases. The Bee Department, Rothamsted Experimental Station, Harpenden, Herts., will help you if you are in any doubt.

If you develop a good strain of bees let your bee-keeping friends know when you have a spare Queen cell or nucleus. Never charge bee-keepers for Queen cells. If you improve the stocks of your neighbouring bee-keepers this will probably in the end help you to maintain your high quality, for who knows but that a drone from a neighbouring bee-keeper may fertilise your next virgin Queen.

Always be willing to look after the stocks of another bee-keeper if he is ill or called away on important business ; do not expect any reward for this, you yourself may need help some day.

If you are called in to remove a swarm from premises near another apiary, take the swarm, but before removing it go to the other bee-keeper to ascertain if he has lost it, and if that be the case the rightful owner should be allowed to hive the swarm. If, however, you are the rightful owner, you should proffer some good evidence that it is your swarm.

There is an ill-founded notion amongst bee-keepers that it is permissible to trespass in pursuit of their swarms. A case was, however, recently decided in the High Court from which it is clear that such notion is now without effect. Keep sweet with your neighbours and you will not meet with difficulty over trespass.

Some bee-keepers in the hope of catching stray swarms

leave exposed an empty hive but fitted with brood-chamber and old combs. This is " not done " amongst good bee-keepers because if a swarm from another beekeeper's hive should emerge and the scouts from the swarm find this desirable home, then the swarm may decamp before the rightful owner has time to hive it.

Manipulation of a Hive

When opening up a hive great care should be taken to ensure that the bees are as little disturbed as is possible. Naturally the apparent breaking up of their home does cause the bees a certain amount of distress and concern and this is shown in two ways. Either the bees will become angry and will fly out and attack the operator or they will make for the honey cells and gorge themselves with honey. The latter they do as a matter of instinct, for they will need that honey if they are compelled to find a new home.

Rough handling of the hive makes bees angry and they will attack the operator.

The illustration, Plate III, shows the author's protection when handling strange or " active " bees.

This method of approach has been found most satisfactory by the writer. First, have everything you intend to use in the hive handy and immediately available. Light the smoker and give the hive three good puffs of smoke direct into the entrance. Then, without hurrying, remove the roof and the outer lifts so that the interior is easily accessible. Do not jolt or jar the hive in any way. Remove the blankets, raise a corner of the quilt and give a small puff of smoke into the exposed combs. Pull off very gently the quilt and usually the bees will have their heads stuck into the honeycomb. Then take a " stink cloth," that is a cloth soaked in a weak solution of carbolic, lysol or Jeyes' fluid, and cover the exposed brood-combs Leave it there for a few seconds and then remove it. The

bees will be completely subdued. Take the hive tool or an old broad chisel and remove the division board by levering up the lugs. Take out the end comb nearest the division board, look and see if the Queen is on the comb. If not, thump the bees off the comb by holding the comb in the left hand by a lug and thumping the left hand sharply with the right fist. The bees may

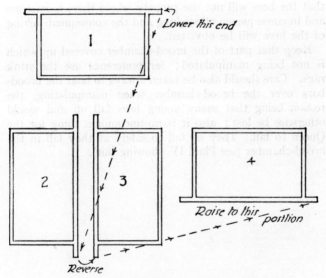

FIG. 31

Inspecting a Brood Comb

either be thumped on to the top of the brood-frame, or on to a board running up to the hive entrance, the latter for preference. Now place this outside comb on one side and lift up each comb in turn (see Plate IV, showing the comb resting across the walls of a lift). If you are in search of Queen cells it will be necessary to go through each comb with great care. There is a right and wrong way of

E

turning combs over. Plate IV shows the correct method of inspecting a brood-comb and (Fig. 31) the correct methods of turning a comb over.

If whilst going through the brood-chamber it is found that the brood-bars stick to the runners in the brood-chamber, then smear the runner with vaseline and note the difference at the next manipulation. It will be found that the bees will not use propolis where there is vaseline and in consequence all sticking and the consequent jarring of the hive will be obviated.

Keep that part of the brood-chamber covered up which is not being manipulated; for preference use the stink rags. Care should also be taken always to hold the brood-bars over the brood-chamber when manipulating, the reason being that many young bees fall off and would otherwise be lost; also it is no uncommon thing for the Queen to fall. They all fall to safety as they fall in the brood-chamber (see Plate IV, showing this).

CHAPTER XII

DISEASES OF BEES

LIKE most other insects bees are liable to suffer from diseases; some are mild and curable whilst others are malignant and deadly.

Bee diseases affect bees in two forms, namely, diseases of adult bees and diseases of brood.

Adult diseases are principally Acarine disease, Amœba disease and Nosema disease. Dysentery, most forms of Paralysis and Poisoning are not diseases but are complaints of the adult bee other than disease.

Amœba disease is very rare and does not seem to have occurred in this country in recent years. Nosema is widespread and fairly abundant and Acarine is both widespread and extremely abundant.

Acarine Disease

This disease in the early part of this century wrought untold havoc amongst the colonies of bees in this country. Bee-keepers became almost desperate about this plague which was alleged to have commenced in the Isle of Wight. Most people who have a scanty association with bee-keeping have heard of this dread disease. But science has mastered it and precautionary methods have been introduced which can now ensure the bee-keeper comparative freedom from this trouble.

The study of the anatomy of the bee will show that like other insects the bees have not got lungs as we understand them. Their breathing is confined to what is known as " tracheæ " which are a series of breathing tubes branching one from the other. The parasite, which is the cause of this disease, is called the " Acarapis Woodi " and from these names the term Acarine disease is derived.

This mite lives in the tracheæ of the thorax of the adult bee and is spread from bee to bee via the female of the species which when pregnant colonises by leaving its host and finding its way to a young bee less than 5 days old. It has been shown that it will find its way into Queen, drone and worker alike.

The female having found a new host proceeds to hatch out her young through the stages of eggs, larvae and adult. It is thought by some that the adults and the larvae feed upon the blood of the bee by sucking it through the tissues of the trachea although this is not yet proved. When they become sufficiently numerous they completely choke up these breathing tubes. This handicaps the bees in their flight and many die off through exhaustion during their absence from the hive and without the knowledge of the bee-keeper. This may show itself either in the fact that the stock fails to produce any stores of honey and also by the fact that the hive appears constantly weak.

The symptoms of the disease are :—

(1) *Crawling*. The bees find that they are too weak to fly and crawl out of the hive, fall to the ground and climb to the top of spikes of grass. This crawling should not be confused with the " dancing " of young bees on the alighting board. Neither should it be confused with an old bee returning after foraging with a heavy load. That bee can easily be distinguished by its shiny body and frayed wings. It is the bees which crawl *out* of the hive and drop off the landing board which give the tell-tale symptom. If there is what is described as mass crawling going on, that is bees crawling all over the ground round the hive then it is almost certain that the hive is badly infected.

(2) *Swollen abdomen*. For some reason said by some to be a form of poison which the mites return to the host when sucking its blood, the abdomen of the bee becomes

grossly distended and this symptom indicates an advanced stage of diarrhœa. This symptom does not always occur in an infected bee.

(3) *Laziness of bees.* If large numbers of adult bees are found loafing about the hive on a day when other stocks are out foraging then the bee-keeper should view it with suspicion and watch carefully for other symptoms. Laziness may also be associated with swarming which may take place within a day or two of laziness being apparent.

(4) *Dislocated wings.* Adult bees with this disease frequently appear crawling on the landing board of the hive unable to fly with their wings stuck out at odd angles.

There are other points of interest which may be given in regard to this disease.

The stocks affected dwindle rapidly and usually die out between one and two months although infected stocks may live for years.

It is thought that the disease is transferred from stock to stock by robber bees from an infected hive passing it on to the clean hives. Likewise " drifting bees," that is bees which enter the wrong hive, the hives being too close together. Natural swarms are a source of suspicion when they come from other places. Keep your eye on them for disease and never manipulate a strange swarm before operating upon " clean bees." Do it after and then wash your hands well.

The mite will not live very long when absent from its host and therefore it is quite safe to use the equipment of an infected hive after a period of say one month.

There are ways of avoiding this disease. The most important is to see that you do not buy bees from strange sources unless you are assured in writing that the apiary is free from disease. Do not place your bees near those of a " dirty " bee-keeper.

If in any doubt about your bees send a sample of them,

E

15 or 20, in a matchbox to the Rothamstead Experimental
Station, Harpenden, Herts.   No charge is made for a test.
This station is very anxious to track down disease and you
will be helping them as well as yourself.

These remarks apply to all bee diseases.

The Treatment of Acarine Disease.

The Frow Treatment.

About fifteen years ago **Mr. R. W. Frow** informed the
bee-keeping world that he had discovered a cure for this
disease.   He had discovered a way of killing the mites
without killing the bees.   Dr. Rennie had already pointed
out this as the way to succeed but to Mr. Frow has gone
the due praise for the boon which his treatment has been
to all bee-keepers.   As a precaution against their colonies
becoming victims serious bee-keepers always apply the
treatment to prevent their stocks from becoming victims
of the disease.   The time may not be far distant when
each bee-keeper will, in the interests of all other bee-
keepers, be required by law to treat his bees annually.
Sheep are dipped in a preventive dip : why not bees
likewise ?   This would almost eradicate the trouble.

The principle of Mr. Frow's treatment is simple.   A
volatile mixture is introduced into the hive and it permeates
every nook and cranny even getting into the tracheae of
the bees.   There it does its deadly work on the live mites.
It does not affect the eggs so that a further dose is necessary
to dispatch them when they hatch out.

The prescription for the mixture which Mr. Frow has
so generously made free to all is as follows :—

      Nitrobenzol or
          Oil of Mirbane   ...   2 parts by volume.
      Safrol Oil   ...   ...   1 part by volume.
      Petrol   ...   ...   2 parts by volume.
Ordinary good quality petrol is all that need be used.

It should be thoroughly understood that this mixture

is poisonous and therefore should be handled with care.

There are two methods of treatment. One by lifting the front of the brood-chamber and slipping underneath a felt pad on which 20 minims of the liquid have been poured, and repeating the dose daily for six days to catch the mites which hatch out late. The other method of application is to put a pad of felt over the feed hole and so introduce the vapour from above. Either method is satisfactory.

Again, there is a difference of opinion as to the dose which should be given. Some say that a single dose of say 60 minims is as good as six doses of 20 minims. Both in practice have been found successful. The large dose seems to cover the hatching out period.

The treatment should be applied during the cold weather, November to February is a good time. The best time for application of the " Frow " or " Modified Frow " treatment has been found by most research laboratories to be in February or March as soon as possible after the bees have had a good cleansing flight. A measuring glass can be purchased from any dealer in bee-keeping appliances or a chemist. Multiple chemists have this treatment already made up with printed instructions as to its use.

The Oil of Wintergreen Treatment.

Experiments have been successfully carried out by introducing oil of wintergreen (methyl salicylate) into the hive. This is placed in a bottle between the combs at the back of the hive. One ounce of oil of wintergreen is placed in a bottle with a wick sticking up between the combs. This slow fumigation carries on until the bottle is empty.

There is at the present time a great shortage of safrol and there is not sufficient to meet the present demands of bee-keepers for treating the increased number of colonies of bees. An admirable substitute has been found in methyl salicylate (oil of wintergreen) for safrol. It has been found that if the proportion of nitrobenzine is increased the results are equally as effective as the original Frow recipe. Further, a substance known as Lingroin, which is a petroleum product, is quite suitable for this mixture.

The recipe for the substitute which is just as effective as the original Frow recipe is as follows :—

Methyl Salicylate
(Oil of Wintergreen) ... 2 parts by volume.
Nitrobenzine ... ... 6 ,, ,,
Lingroin ... ... 5 ,, ,,

Method of Application.

A pad of felt or flannel should be placed over the feed hole of the quilt covering the brood-chamber. If a crown board is used then over the feed hole in that board.

Seven doses of 30 minims (half a drachm) should be poured drop by drop over the pad. The doses should be administered alternate days taking a fortnight in all. But it is desirable to leave the pad on for a further two or three days. The pad should then be removed and the feed hole covered. The treatment is then ended.

Time of Application.

During February and March the bees will take their first cleansing flights. It is after these flights that the treatment should be given. If, however, it is known in the Autumn that the hive is infected the oil of wintergreen treatment described earlier should be administered during the winter.

If the beginner is in any doubt he should call in the local expert who will increase the dosage if that is found necessary.

Nosema Disease

This is not a very common disease in this country. It

is caused by an animal parasite which attacks the stomach of the bee called the chyle stomach. The parasite is called the " Nosema Apis " hence the term Nosema disease. It attacks the stomach lining, which it destroys, and the bee dies. The bees which are about to die resemble bees suffering from acarine disease. They crawl because they have not the strength to fly, but they fall off the landing board because they have no energy. They fall on their backs and die.

The disease attacks the workers, drones and the Queen. If the Queen is attacked it is doubtful if the stock will survive. But if the disease attacks the hive at the height of the season the rapid breeding will exceed the losses. The effect will be that it will remain a weak stock and the stock may recover. However, the losses will make the stock a poor honey producer.

The remedies for this disease.

All dead bees should be collected and burnt. The bees in severe cases should also be killed to prevent the spread of infection, the combs should be burnt, the hive scorched with a blow-lamp and the ground round the hive dug over and sprinkled with quicklime. This is to destroy the spores which are present in the discharged faeces of the infected bees.

Prevention of the disease may be effected by keeping all hives clean and the bees strong in numbers. Their drinking water should be watched. Queens and new stocks should only be obtained from *bona fide* sources.

Dysentery

This complaint which is not of itself a disease causes the bees to deposit their excrement within the hive. The excrement is dark in colour and cloudy, and offensive in smell. In the hive it is smeared on the combs, the walls and the floor. It is caused mainly by the bee-keeper through carelessness, and can be avoided. It is caused through incorrect feeding; feeding with glucose or brown sugar. Feeding too late in the Autumn so that the bees do not seal their stores, allows the unsealed stores to ferment and fermented honey gives the bees dysentery. Honeydew and honey containing too much dextrin will also cause this disease. If the hive is damp this is conducive to dysentery, particularly in the winter months. Sometimes, too, if the bees are confined to the hive they will develop this disease.

The treatment is to remove the soiled combs and to replace them with sealed stores, placing these in a clean brood-chamber. A cake of candy should be placed over the feed hole and the bees will soon recover.

It is wise for the bee-keeper to keep a comb or two of sealed natural stores on one side in case this disease should have to be dealt with. The hive should be kept warm and the bees left undisturbed.

Paralysis

Little is known of this disease. The symptoms are, first, the bees affected are shiny and hairless. When the dead bees are squeezed a watery liquid comes out. This has a bad smell. The live bees will be seen dragging out the dying bees. The stock may die out entirely, especially if the Queen is affected.

The remedies suggested are first raise the stock of bees higher from the ground so that affected bees cannot easily regain admission to the hive. Secondly, requeen the hive. Dusting the interior of the hive with flowers of sulphur and feeding the bees with syrup treated with thymol, as recommended by some, is not advised, as these substances may do more harm than good.

Poisoning

This is not really a disease of bees but this appears to be a suitable place to mention it. In fruit-growing districts when apple blossoms in particular are in bloom, it is the practice of the fruit-grower to spray his trees with certain arsenical preparations in order to deal with the pests on the trees. The bees gather the nectar containing the poison and return to the hive. They sometimes, too, gather water in the form of dew from the grass under the sprayed trees and this also is contaminated. Most of the bees will die before they reach the hive but they may return with pollen contaminated by the poison in which case other bees will be affected too. This, too, is the only known way in which brood becomes poisoned. The bee-keeper should tactfully approach the fruit-grower before the trees come into bloom and point out that the advice of the Ministry of Agriculture is that spraying before blossom is more effective than when the blossom is out. That is the only safe way to prevent what would otherwise cause " spring dwindling " among the bees.

Brood Diseases

The principal brood disease is Foul Brood. There are two distinct diseases, American Foul Brood, known as A.F.B., and European Foul Brood, known as E.F.B., each caused by different organisms.

American Foul Brood is unfortunately very prevalent in this country and every bee-keeper should do his best to take part in the campaign to eradicate it. The government have already taken steps to assist in its extermination by making an order imposing penalties for failing to deal with the complaint.

The cause of American Foul Brood is an organism known as " Bacillus Larvae " and is found in larvae affected by the disease.

The bee-keeper should constantly be on the watch for this disease. Every time a hive is examined a careful watch should be made. Look for the patches of brood and if there are a number of misses in the regular cappings look and see what is in the uncapped cells. Sealed brood usually occurs in blocks and a good number of uncapped cells give the first clue. Then see if any of the cell cappings are concave instead of the healthy convex cappings. Many of the cappings may be partially removed by the bees in their effort to get rid of the foetid contents of the cells. A recognised symptom is the foul smell emitted from the decaying larvae. Smell is not a safe guide as to the presence of foul brood. Some smell may or may not be present. In the case of A.F.B., the brood does not die until after it has been capped. In the case of E.F.B. the larvae are affected and die prior to sealing. Infected larvae have not got the pearly whiteness of healthy larvae. They appear flabby, yellowish brown and then turn slimey.

The disease is spread from hive to hive through robber bees and drifting bees. It can generally be prevented by the bee-keeper taking the following advice.

Always purchase bees from a known breeder of repute who can certify his apiary free from the disease.

Never feed your bees on "strange" honey.

Never purchase old combs for use in your apiary.

Never interchange combs in "doubtful" hives.

Always disinfect borrowed equipment before use and always disinfect it before returning it.

Always wash your hands after "handling" another's bees.

The safest treatment is destruction of the bees, combs and quilts and the scorching of the brood-chamber.

Neither A.F.B. nor E.F.B. can be cured by the use of any known drug or disinfectant.

The beginner should call in the expert whose name he can readily get from the local association's secretary.

He may advise two alternative methods of saving the stock, i.e., by shaking or by the artificial swarming method. But beginner, please do not try these cures without expert advice.

If the beginner will join the local association he will

enjoy the benefit of insurance against this disease and will not be the loser by the loss of the bees. The local association can obtain this cover by being affiliated to Bee Diseases Insurances Limited.

Remember always that if you do not deal with this disease you are a danger to the rest of the bee-keepers in your district. If in doubt call in the expert.

European Foul Brood

This disease is not so prevalent as the American variety. Its origin has now been definitely proved, the causal organism has recently been successfully cultured on an artificial medium. " Bacillus Pluton " is the cause of the trouble. Fortunately it is not so widespread as American Foul Brood. Although it is often present a foul smell is not necessarily associated with this disease. It all depends what secondary organisms are present in addition to the causal organism. The larvae die before they reach the pupa stage and emit a foul stench. The bees succeed in clearing out the dead larvae. In the case of American Foul Brood the larvae die after they reach the pupa stage and are capped over and the bees do not succeed in removing the dead in A.F.B. An excellent way of testing the difference between the two diseases is to take a matchstick and twist it in the dead matter in a cell. In the case of A.F.B., the matter will pull out in what is often described as ropiness, whereas with E.F.B., the matchstick comes away readily without pulling the dead matter out in a thin thread.

With A.F.B. the symptoms can be found at any time during the breeding season, whereas with E.F.B. the symptoms are usually found in the Spring but disappear as the season progresses.

Strong stocks always seem to be the best defence against bee diseases but unfortunately strong stocks are the very ones that are most likely to contract foul brood if it is present in the neighbourhood, as these are the colonies which, while being unlikely to be robbed on account of their strength, are most likely to rob out colonies weakened by foul brood.

Refrain from using borrowed equipment and do not buy bees in a doubtful market.

Cure for the disease has been professed by many but the beginner again is advised to call in the expert for advice.

In the absence of such advice destruction of the stock is the only way to eliminate the disease.

No cure or method of eliminating European Foul Brood from a colony of bees is known. Many laboratories have been attempting to find such a method for years and have also tested all methods suggested, all with complete lack of success. Total destruction is the only " cure " for European Foul Brood.

Addled Brood

This is a disease associated with the Queen and can readily be overcome by removing the Queen and replacing her with another Queen. The disease then disappears.

The symptoms are briefly as follows. The larvae usually, but not always, die in or before the pupal stage. That is after the cells have been sealed. This is one way of distinguishing it from foul brood. The cappings sink and the larvae have a shrivelled appearance as if ill-nourished and have a pungent acid smell which is unpleasant. The bees uncap these dead pupae and attempt to drag them out. Occasionally dead pupae are found on the alighting board outside the hive. These should be collected and sent to the Rothamsted Experimental Station for verification together with a complete comb. Sending a complete comb assists materially in the diagnosis of the complaint.

Chalk Brood

This disease is caused through a mildew or fungus.

Sealed brood only is affected. It is supposed to be more prevalent among the black varieties than amongst the yellow Italian varieties, although the writer has experienced it in a yellow stock. Curiously the drones are affected more than workers. The sealed larvae dry up and have the appearance of greyish mummified corpses. These are often seen near the entrance to the hive. This is not a serious disease and some well-known text-books say that it can usually be cleared by requeening, preferably with one of the yellow variety, but there is really no need to requeen at all. The colony if kept strong will probably overcome this disease itself in mild cases. In severe cases the bees should be transferred to a clean hive with new combs or foundation.

Chilled Brood

In addition to the troubles with brood caused through infection there are troubles caused through accident or through carelessness of the bee-keeper.

Manipulation of the hive too early in the season when the days are sunny but cold. This chills off the brood and the bees drag out the dead grubs and sometimes the dead sealed larvae. This puts the stock backward.

Similarly if the brood-nest is spread too rapidly there may not be enough bees to protect the brood and in consequence chilling occurs.

Starved Brood

In the early Spring when the brood nest is expanding a cold snap may come along which prevents the bees foraging. They will then use up stores and unless there are sufficient of these or candy or syrup provided the brood will die through lack of food. The bees then drag the corpses out and there is likelihood of the whole stock dying out. This may occur in the Autumn in which case the bees after dragging out their dead may leave the hive as a hunger swarm. Swarms in August onwards should always be suspected as being hunger swarms and taken with necessary caution.

Enemies of Bees

There are several enemies of bees each of which affect the stock in different ways.

The toad may sit on the landing board of the hive and pick up the bees as they land with their load. But they also pick up dead and perhaps diseased bees lying round the entrance so should not be discouraged.

Mice occasionally enter a hive in the winter and cause havoc. They can be prevented by a strip of Queen excluder being placed over the entrance.

Tits frequently tap at the entrance to hives in the winter time. The bees which are attracted to the entrance are picked off one by one by the tits.

Ants collect in the quilt of the hive and are a nuisance. They can be prevented from doing this if the four legs of the hive are stood in tins of old motor sump oil.

The same applies to earwigs although they can and will fly to the hive so that placing the hive legs in oil may not prevent all of these insects from getting into the hive.

The worst enemy to the hive is the wax moth. There are two species and they may do irreparable damage. They lay eggs in the comb and the grubs will eat very rapidly through a comb, soiling it with faeces. Great care should be taken with the storage of combs when they are not in use.

Hornets in other countries are a menace to bees. The author has seen in Palestine Arabs—who produce most of Palestine's honey—standing with large swatters by beehives swatting hornets in large numbers. Hornets even keep the bees indoors during the mid-day hours.

The best known method of fumigation to prevent wax

moth is the Paradichlora-benzine, commonly called " P.D.B." and it is used in the following way :—

First, stack the brood-chambers and shallow frame supers complete with the combs to be treated one above the other. The space where the chambers meet should be sealed over on the outside with strips of gummed paper so that the effect is to allow a through draught of air. The lowest brood-chamber should be placed on a sheet of newspaper which should be pasted up on the outer sides of the brood-chamber.

Secondly, place an ounce or two of these crystals on a saucer or plate on the top of the shallow frames in the top crate, and cover the whole by pasting a sheet of newspaper over the top, or by using a national roof as a cover.

P.D.B. crystals give off heavy volatile fumes and these fumes will penetrate throughout the whole stack of brood-chambers and crates.

If this is done in the Autumn and the P.D.B. crystals replenished in January or February the combs will be kept proof against wax moth.

Before using the brood-chambers or shallow bars, air them in the open air for a few hours to get rid of the P.D.B. fumes.

It is particularly important to note that in no circumstances should P.D.B. be used direct in the hive, otherwise fatal results will follow.

APPENDIX

BOOKS OF REFERENCE

The reader is advised to take advantage of the opportunities of reading as much as he can about this absorbing subject and the following list of books is recommended for selection.

Author.	*Title.*
APIS CLUB	Diseases of Bees.
BOFF, C.	How to Grow and Produce Your Own Food.
CHESHIRE, F. R.	Bees and Bee-keeping.
COWAN, T. W.	The Honey-bee.
COWAN, T. W.	British Bee-keepers' Guide Book to the Management of Bees in Movable-comb Hives, and the use of the Extractor.
DADANT, C. P.	Dadant System of Bee-keeping.
DIGGES, J. G.	Practical Bee Guide.
EDWARDES, T.	Lore of the Honey-bee.
FABRE, J. H.	Bramble-bees, and Others.
FRANÇON, J.	The Mind of the Bee.
GILMAN, A.	Practical Bee-breeding.
HASLUCK, P. N. (Ed.)	Bee hives and Bee-keepers' Appliances.
HEMPSALL, W. H.	Bee-keeping, New and Old.
HOOPER, M. M.	Commonsense Bee-keeping.
LAWSON, J. A.	Honeycraft in Theory and Practice.
MACE, H.	Book about the Bee.
MACE, H.	Modern Bee-keeping.
MACFIE, D. T.	Practical Bee-keeping and Honey Production.
MAETERLINCK, M.	Life of the Bee.
MANLEY, R. O. B.	Honey Production in the British Isles.

PELLETT, F. C. . . Productive Bee-keeping.

RENDL, G. . . . Way of a Bee.

ROOT, A. I. . . ABC and XYZ of Bee Culture.

SNELGROVE . . Swarm Control.

STEP, E. . . . Bees, Wasps, Ants and Allied Insects.

WEDMORE, E. B. . Manual of Bee-keeping for English-Speaking
    Bee-keepers.

WILLIAMS, C. . Story of the Hive.

INDEX

B